About Island Press

Since 1984, the nonprofit Island Press has been stimulating, shaping, and communicating the ideas that are essential for solving environmental problems worldwide. With more than 800 titles in print and some 40 new releases each year, we are the nation's leading publisher on environmental issues. We identify innovative thinkers and emerging trends in the environmental field. We work with world-renowned experts and authors to develop cross-disciplinary solutions to environmental challenges.

Island Press designs and implements coordinated book publication campaigns in order to communicate our critical messages in print, in person, and online using the latest technologies, programs, and the media. Our goal: to reach targeted audiences—scientists, policymakers, environmental advocates, the media, and concerned citizens—who can and will take action to protect the plants and animals that enrich our world, the ecosystems we need to survive, the water we drink, and the air we breathe.

Island Press gratefully acknowledges the support of its work by the Agua Fund, Inc., Annenberg Foundation, The Christensen Fund, The Nathan Cummings Foundation, The Geraldine R. Dodge Foundation, Doris Duke Charitable Foundation, The Educational Foundation of America, Betsy and Jesse Fink Foundation, The William and Flora Hewlett Foundation, The Kendeda Fund, The Andrew W. Mellon Foundation, The Curtis and Edith Munson Foundation, Oak Foundation, The Overbrook Foundation, the David and Lucile Packard Foundation, The Summit Fund of Washington, Trust for Architectural Easements, Wallace Global Fund, The Winslow Foundation, and other generous donors.

The opinions expressed in this book are those of the author(s) and do not necessarily reflect the views of our donors.

THE REBIRTH OF ENVIRONMENTALISM

THE REBIRTH OF ENVIRONMENTALISM

*Grassroots Activism from
the Spotted Owl to the Polar Bear*

Douglas Bevington

WASHINGTON | COVELO | LONDON

Library of Congress Cataloging-in-Publication Data

Bevington, Douglas, 1970–
The rebirth of environmentalism : grassroots activism from the spotted
owl to the polar bear / Douglas Bevington.
p. cm.
Includes bibliographical references and index.
ISBN-13: 978-1-59726-655-0 (cloth: alk. paper)
ISBN-10: 1-59726-655-8 (cloth: alk. paper)
ISBN-13: 978-1-59726-656-7 (pbk.: alk. paper)
ISBN-10: 1-59726-656-6 (pbk.: alk. paper)

1. Environmentalism—United States. 2. Biodiversity—United States. 3.
Environmental policy—United States. I. Title.
GE197.B48 2009333.95'160973—dc22

2009006757

Printed on recycled, acid-free paper

Design by Joan Wolbier

Manufactured in the United States of America

CONTENTS

PREFACE AND
ACKNOWLEDGMENTS

This book developed out of research I undertook while completing my PhD in sociology at the University of California, Santa Cruz. For that study, I conducted interviews with sixty-two participants in biodiversity protection campaigns: Peter Bahouth, Kathy Bailey, Michael Bean, Jim Bensman, Marty Bergoffen, Bryan Bird, Monica Bond, Darryl Cherney, Jamie Rappaport Clark, Susan Clark, Kate Crockett, Brendan Cummings, John Davis, Bill Devall, Michael Dorsey, Cynthia Elkins, Brock Evans, Jennifer Ferenstein, Peter Galvin, Eric Glitzenstein, Chelsea Gwyther, Dan Hamburg, Keith Hammer, Chad Hanson, Connie Hanson, Tim Hermach, David Hogan, Shane Jimerfield, Jim Jontz, Tracy Katelman, Josh Kaufman, Matthew Koehler, Cecelia Lanman, Andy Mahler, Julie Miller, Ken Miller, Sue Moloney, Ned Mudd, Brian Nowicki, David Orr, Michael Passoff, Daniel Patterson, Karen Pickett, Jill Ratner, Mike Roselle, Jeanette Russell, Todd Schulke, Michael Shellenberger, Kassie Siegel, Robin Silver, Rhiwena Slack, Bill Snape, Karyn Strickler, Kieran Suckling, Charlotte Talberth, John Talberth, Doug Tompkins, Jay Tutchton, Tom Van Dyck, Rene Voss, Tom Woodbury, and Margaret Hays Young. These interviews were conducted primarily between December 2004 and August 2006. The quotes in this book come from these interviews unless otherwise cited. I would like to thank all of the interviewees for their time and candor. Many of them were also very helpful in answering follow-up questions, providing access to their archives, and hosting me during my travels. This project would not have been possible without them.

I should also note that there were four interviews that I solicited but was unable to conduct. Carl Pope of the Sierra Club declined my request, citing a busy schedule. Jasper Carlton of Biodiversity Legal Foundation was unable to be interviewed due to health considerations. Steve Kallick of Pew Charitable Trusts was unable to get permission from his organization to speak with me and recommended that I instead talk with Joshua Reichert of Pew. However, I did not receive a reply to my interview request from Mr. Reichert.

My research was informed not only by the interviews, but also by direct experience in biodiversity activism. Over the past two decades, I have been involved in forest and endangered species protection campaigns from a variety of positions and perspectives. I have worked with national environmental organizations (the Sierra Club and National Audubon Society) and grassroots groups (the John Muir Project). I have participated at the local level (the Bay Area Coalition for Headwaters) and in Washington, DC (the Endangered Species Coalition). I have been involved with legislation (the National Forest Protection and Restoration Act), litigation (as a plaintiff and standing declarant for endangered species protection lawsuits filed by the Center for Biological Diversity), and direct action (the Headwaters campaign). And I have been employed in fundraising (Greenpeace) and as a funder (Foundation for Deep Ecology). In contrast to some academics who emphasize detachment from the subjects of their study, I have found that social movement researchers can create more useful scholarship through direct engagement with social movements. (For a discussion of this methodology, see Bevington and Dixon, "Movement-Relevant Theory.")

I am very grateful to my dissertation committee—Andrew Szasz, Barbara Epstein, and Richard Flacks—for supporting this approach and guiding me through the dissertation-writing process. I would also like to offer special thanks to Chris Dixon, who served as a virtual fourth committee member, providing encouragement and extensive feedback on all of the chapters. And I received helpful suggestions on portions of the dissertation from Gary Bevington, Tim Ingalsbee, and Mark Lovelace.

After I completed my dissertation, I was offered the opportunity to direct the forest protection program of Environment Now, a grantmaking foundation with a particularly strong commitment to grassroots groups. I am very grateful to the board and staff of Environment Now for giving me this opportunity to apply some of the lessons I learned during my research. I am also honored to have recently joined the board of directors of the Fund for Wild Nature, which transfers the donations from its members into grants for grassroots biodiversity protection groups.

I would also like to offer my thanks to the staff of Island Press for guiding me through the process of converting my manuscript into book form.

Finally and foremost, I want to express my heartfelt gratitude to Shaye Wolf for her guidance and encouragement throughout this entire process. When I first met Shaye, we were both in graduate school. She was actively involved in protecting endangered species as a conservation biologist, but she had not heard of the Center for Biological Diversity prior to meeting me. I shared my dissertation chapter on the Center with her as I wrote it. After reading it, she was inspired to apply for a position with the Center. She is now a staff biologist for Center for Biological Diversity, working to protect imperiled wildlife from the effects of global warming. I hope that the examples of effective environmental activism collected in this book will offer similar inspiration for other readers as well.

1

The Rise of Grassroots Biodiversity Activism and the Rebirth of Environmentalism

On a sunny day in July 1987, a rusty converted fishing boat called the *Divine Wind* set out from Seattle with a crew of twenty-one environmental activists on board. Their goal was to find a drift-netting vessel on the high seas and ram it. The hull of their boat had been deliberately reinforced for this purpose. The activists called themselves the Sea Shepherd Conservation Society. Sea Shepherd was a group that, like its land-based counterpart Earth First!, would intervene directly to stop environmentally damaging activities, a tactic known as "direct action." Both groups would sometimes damage property as part of their direct action, but they took great care not to injure other people, even during dramatic acts such as ramming a ship. Sea Shepherd had used this tactic previously in its campaign to stop ships involved in illegal whaling. In 1987, the group hoped to use the same tactic to deter ships engaged in high seas drift-netting and draw public attention to this particularly destructive form of fishing. Drift-netting entails using massive nets—often thirty to forty miles long—which, in addition to catching fish, would also trap and drown more than a million dolphins, whales, sea turtles, seabirds, and other marine wildlife each year as bycatch.

One member of the Sea Shepherd crew that year was Brendan Cummings, a young environmental activist who took a leave from college to volunteer for the drift-net campaign. Sea Shepherd needed an engineer, and although Cummings's only previous experience was repairing his own motorcycle, he was assigned the task of keeping the *Divine Wind*'s decrepit engine running throughout the campaign. During the summer of 1987, Sea Shepherd's crew pursued the drift-netting fleet across the northern Pacific Ocean. Ultimately, the fleet eluded them that year, though during a subsequent campaign in 1990, Sea Shepherd rammed two drift-net ships, disabling their net deployment equipment.[1]

In the meantime, Brendan Cummings had returned to college. He remained involved in environmental activism and attended law school in the mid-1990s. Afterward he became an attorney for a small environmental group called the Center for Biological Diversity. Like Cummings, the Center's founders had roots in direct action activism, but the Center was known for its extensive use of litigation to enforce environmental laws to protect imperiled wildlife. Cummings would develop a program to apply the Center's tactics to the protection of ocean-dwelling species. Although high seas drift-netting had been banned by the United Nations in the 1990s, he discovered that a smaller form of drift-netting called drift-gillnetting was still taking place in the waters off California. This fishery was incidentally catching and killing leatherback sea turtles, the largest and most imperiled species of sea turtle. Cummings devised a lawsuit under the citizen enforcement provisions of the Endangered Species Act that resulted in seasonal closures of the gillnet fishery along most of California's coast. And in the decade since that lawsuit was filed, not a single leatherback turtle has been killed by the fishery.

Litigation proved to be a powerful tactic for accomplishing the wildlife protection goals that Cummings had first sought to achieve through direct action with Sea Shepherd. The Center for Biological Diversity, which fostered Cummings' litigation, was an example of a new form of environmental activism that emerged in the late 1980s and 1990s—the grassroots biodiversity group.

Grassroots Biodiversity Groups and the U.S. Environmental Movement

Grassroots biodiversity groups have been unsung heroes of American environmentalism during the past twenty years. These are small, radical environmental organizations that protect imperiled wildlife and forests, particularly through aggressive use of litigation. There was a remarkable proliferation of these groups in the late 1980s and 1990s. The new groups were directly responsible for an unprecedented increase in biodiversity protection during that period. For example, by the early 2000s, logging on national forests had plummeted to its lowest level since the 1930s in response to appeals and lawsuits coming largely from these groups. The new groups were able to accomplish a tremendous amount with very few resources. For example, the Center for Biological Diversity was described in its early years as "a handful of hippies operating off of unemployment checks."[2] Yet by 2004, the Center was responsible for getting 335 species of animals and plants protected under the Endangered Species Act, more than any other environmental group of any size.[3] Today the Center is best known for applying the Endangered Species Act to protect polar bears from the effects of global warming.

Despite all of their accomplishments, the grassroots biodiversity groups have largely been overlooked in the recent histories of the U.S. environmental movement. To understand the role of these groups, it is important to situate them in relation to the concept of biodiversity, the laws protecting biodiversity, and the large national environmental organizations that advocate for biodiversity.

"Biodiversity" was a term popularized by biologists in the 1980s in response to growing concerns that human activities were causing other species of life to go extinct at an unprecedented rate.[4] The term served to convey the importance of maintaining a full range of animal and plant species in the face of this extinction crisis. From this perspective, wildlife and forests were not simply scenic amenities for people involved in outdoor recreation, but instead were integral to the "web of life" of healthy ecosystems. Some advocates for biodiversity protection highlighted the value of maintaining healthy and complete ecosystems for their benefits to humans, both in terms of the ecosystem services they provide (such as purifying air and water) and, more fundamentally, as the life-support system for the Earth.[5] Others asserted that all species have an inherent right to exist and that the extinction of any species is an irreparable loss, comparable to genocide.[6] (Indeed, species extinction is sometimes called "ecocide."[7]) From either perspective, it was imperative to protect biodiversity.

Even before the popularization of that term, the 1970s saw the passage of powerful new laws to protect wildlife and forests, such as the Endangered Species Act, Marine Mammal Protection Act, and National Forest Management Act. (For an overview of these laws, see the appendix.) However the federal agencies charged with implementing these laws—particularly the U.S. Fish and Wildlife Service and the U.S. Forest Service—were often reluctant to do so. Enforcement of biodiversity protection laws was especially difficult because the animals and plants that would benefit from that protection did not vote in elections, whereas the businesses that profited from lax enforcement, such as timber corporations and developers, were frequent contributors to the electoral campaigns of the politicians controlling the purse strings for those federal agencies.[8] Thus, the amount of logging on national forests allowed by the Forest Service grew rapidly even after the passage of National Forest Management Act, and imperiled species continued to decline toward extinction as they waited for the Fish and Wildlife Service to protect them under the Endangered Species Act.

Although these laws contained provisions that increased public oversight and enabled the public to sue federal agencies to ensure that they enforced these laws, the large national environmental advocacy organizations were hesitant to use these tools vigorously. To understand the causes of their relative inaction, it is helpful to situate those groups in the context of the developments within the environmental movement following the first Earth Day celebration in 1970. National environmental organizations, such as The Wilderness Society, Sierra Club, and Environmental Defense Fund, grew quickly in the two decades following the first Earth Day. During this time, these organizations went through a process of institutionalization into the policy-making apparatus in Washington, DC. They became political "insiders."

While these changes enhanced the stature of the national environmental organizations (known among environmental activists simply as "the nationals"), a growing number of critics expressed concern that their effectiveness was being undermined in the process.[9] The institutionalization of the nationals tied them to a process of deal-making that would sacrifice some biodiversity protection in order to broker political compromises. However, biodiversity issues are often ill-suited to the policies that are created through such compromises. If, for example, the timber industry seeks to log the last remaining habitat of an imperiled species of owl, a policy decision based on compromise might produce a resolution in which half of the remaining habitat will be protected and in exchange the other half will be logged. From a political perspective this outcome might seem like a fair compromise, but from an ecological perspective it could be a disaster. If that species of owl needs more than half of its remaining habitat in order to survive in the long term, the eventual result of that compromise will be extinction. Investigative journalist Mark Dowie concluded, "Compromise, which had produced some limited gains for the movement in the 1970s, in the 1980s became the habitual response of the environment establishment. It is still applied almost reflexively, even in the face of irreversible degradations. These compromises have pushed a once-effective movement to the brink of irrelevance."[10]

In the realm of forest and wildlife protection, Earth First! emerged as the main alternative to the national environmental organizations in the 1980s.[11] Earth First!ers deliberately set themselves apart from the nationals with their slogan, "No compromise in defense of mother earth." Earth First! did not seek to engage in political deal-making and instead used direct action, including sabotage of environmentally destructive machinery, as its main tactic. It also eschewed formal organization and conventional fundraising. While Earth First! attracted much media coverage and helped to reframe debates around environmental protection, its choice of tactics gave it little direct leverage over federal environmental policy.

Thus, in the late 1980s, biodiversity activists faced a dilemma. They had a choice between two approaches to biodiversity protection, but each of these routes had notable shortcomings. The national environmental organizations offered political access constrained by compromise, while Earth First! represented an unconstrained approach that was not directly influential. Then a new set of environmental organizations emerged around this time to challenge that dichotomy. Much as the mythical sailors in Homer's *Odyssey* had to chart a course between the twin perils of the jagged rocks of Scylla and the whirlpool Charybdis, the new groups would seek to find a third path that avoided the limited influence of Earth First!'s direct action tactics on the one hand without being pulled into the political constraints of the nationals on the other.

Some activists coming out of Earth First! and others inspired in part by Earth First!'s "no compromise" attitude formed new groups with names such as the Biodiversity Legal Foundation, Center for Biological Diversity, Forest Conservation Council, Forest Guardians, Heartwood, John Muir Project, Native Forest Network, Southern Appalachian Biodiversity Project, and Wild Alabama. While some of these groups focused on the protection of forests and others specialized in using the Endangered Species Act to protect wildlife, there was considerable overlap in these activities, and all of the groups shared an overarching goal of preserving biodiversity. Unlike Earth First!, these new groups were small formal organizations that used only lawful tactics, particularly litigation. At the same time, these biodiversity protection groups were notably different from the national environmental organizations. Indeed, these groups identified themselves as "grassroots" to distinguish themselves from the nationals, although that term did not mean that their efforts were narrowly local in scope. Because biodiversity protection involved federal lands such as the national forests and federal laws such as the Endangered Species Act, the grassroots groups were primarily focused on transforming the policies of the federal government.

In contrast to the national organizations, many of the founders of the new groups described themselves as radicals. The word "radical" has many connotations.[12] For these activists, the label could be best defined as "unconstrained," in the sense that they were willing to pursue the full environmental protections needed to preserve species from extinction even when those goals were considered controversial or politically unrealistic. While no group can be absolutely unconstrained, the grassroots biodiversity groups were qualitatively different from the moderate national environmental organizations.[13] Unlike the nationals, the grassroots groups did not focus their efforts in Washington, DC, and were often critical of the political compromises being brokered there by the nationals. They were not political insiders, nor did most aspire to be. As a result, the new groups were not constrained from filing lawsuits against the federal government for its failure to enforce its own environmental laws in cases that the national organizations avoided as too politically controversial. The grassroots biodiversity groups were therefore able to apply these laws much more extensively than had been done up to this time. Using litigation, they filled a wide gap in the enforcement of environmental laws that had been left by the political constraints on both the federal regulatory agencies and the national environmental organizations. Consequently, these grassroots groups had an unprecedented impact on the implementation of federal biodiversity protection policies throughout the 1990s and into the 2000s.

Three Grassroots Biodiversity Protection Campaigns

The accomplishments of the grassroots biodiversity groups were not easily achieved. These groups were small in terms of their organizational size—often fewer than half a dozen staff members—and their very limited budgets, yet they had to operate in a field that was already dominated by large national environmental organizations with a long history of working on forest and wildlife issues. Indeed, the role played by the nationals created one of the biggest challenges for the new grassroots groups.

To illustrate the challenges and accomplishments of the grassroots biodiversity groups, I focus on case studies of three campaigns that exemplify not only the types of issues covered by these groups, but also three distinct approaches that they took in their relationships to the dominant national organizations, summarized here as (1) never mind the nationals, (2) transform a national, or (3) become a national.

HEADWATERS FORESTS CAMPAIGN
Issue: *Forest protection on private lands*
Relationship to the nationals: *Never mind the nationals*

The Headwaters Forest campaign sought to prevent the logging of the last ancient redwood trees on lands controlled by the Maxxam corporation. Many environmental organizations were reluctant to get involved with forest protection on private lands, but a network of bold grassroots groups emerged to defend Headwaters Forest. One group, the Environmental Protection Information Center, was particularly effective in using litigation to block Maxxam's logging. At the same time, there was a proliferation of new small groups in the region that experimented with a variety of confrontational tactics to put pressure on Maxxam. The Headwaters campaign went on for many years with little or no participation by the national environmental organizations. Undeterred by the absence of the nationals, the grassroots groups used informal coordination and later created a more formal coalition as an alternate means to magnify their influence. In response to this campaign, the government ultimately intervened to purchase and protect part of Headwaters Forest. (See chapter 3.)

ZERO-CUT CAMPAIGN
Issue: *Ending logging on national forests*
Relationship to the nationals: *Transform a national*

In the late 1980s, forest protection activists began calling for an end to all logging on national forests, a position known as "zero cut." However, the national environmental organizations rejected this goal as being too controversial and they hindered efforts by grassroots activists to introduce legislation to achieve it. In response, the grassroots activists undertook a two-pronged campaign to stop public lands logging. On the one hand, a loose network of small forest protection groups used appeals and litigation to stop individual logging projects, leading to a steep drop in logging levels on national forests. Simultaneously, a network of grassroots activists within the Sierra Club, known as the John Muir Sierrans, mobilized the democratic decision-making processes within that organization to elect their own slate of candidates to the board of directors and change the Club's position to support zero cut. They were then able to introduce legislation that called for an end to logging on national forests. (See chapter 4.)

GRASSROOTS ENDANGERED SPECIES LITIGATION CAMPAIGN
Issue: *Protecting wildlife under the Endangered Species Act*
Relationship to the nationals: *Become a national*

The Endangered Species Act provided environmental activists with one of their strongest legal tools for protecting biodiversity, not only in forests but in a wide range of other ecosystems as well. However, in order to be eligible for the ESA's protections, a species first had to be officially added to the government's list of threatened and endangered species. By the early 1990s, most imperiled species were still waiting to receive official protection. In response to the backlog, grassroots biodiversity groups began using petitions and litigation to compel the federal government to list these species. The most successful of these groups was the Center for Biological Diversity. By 2004, the Center had managed to get 335 species protected under the Endangered Species Act, more than any other group of any size. At the same time, the Center was also one of the most successful grassroots groups in terms of its organizational growth. While it started from humble origins, initially operating out of "a shack in New Mexico, its fax machine powered by the sun," by 2004 the Center had a staff of more than thirty people, offices in six states, and a multimillion dollar budget.[14] It was on the cusp of becoming a national organization in terms of its size and resources, while still retaining the bold approach of a grassroots group. (See chapter 5.)

Elements of Success

It is surprising that the grassroots biodiversity groups have been so successful in transforming federal environmental policies. Usually smaller and more radical groups are consigned to the margins of a social movement. Such groups are often seen as lacking influence compared to the bigger and more moderate social movement organizations. Yet over the past two decades, small, bold groups with very limited resources played a central role in many of the most important accomplishments in biodiversity protection. To understand how the grassroots biodiversity groups were able to have such a big impact, I examine these groups in relation to six factors highlighted by social movement researchers: strategy, tactics, organization, funding, movement culture, and political conditions. In so doing, I consider how the choices that biodiversity activists made in relation to each of these factors contributed to and also constrained their effectiveness.

Strategy

Strategy describes a social movement organization's overall approach to making social change. I distinguish between two overarching approaches: an "insider" strategy and an "outsider" strategy. An organization following an insider strategy attempts to make change through conventional forms of participation in electoral politics. The insider depends on privileged access to politicians in order to influence the political deal-making. That access is seen as the primary measure of the insider's power. The insider strategy is commonly used by most interest groups in Washington, DC.

In contrast, social movement organizations that pursue an outsider strategy see their power as primarily coming from outside of Washington. The outsider strategy encompasses the various approaches that do not rely on appealing to politicians to make social change for them. Instead, outsider groups turn to more contentious forms of action that directly challenge harmful activities.[15]

The strategy of a social movement organization has a strong influence on which tactics it uses. Groups using an insider strategy seek to preserve their privileged access to decision makers by not appearing too controversial, which can in turn limit the tactics they use and how they apply them. In contrast, groups using an outsider strategy are relatively unconcerned about their political access and thus freer to pursue more contentious tactics than insiders. A central distinction between the grassroots biodiversity groups and the national environmental organizations is that the former generally followed an outsider strategy, whereas the latter relied on an insider strategy.[16] Those differing strategies in turn shaped the use of litigation as a tactic.

Tactics

Tactics provide the means by which social movement organizations exert influence. One key tactic for biodiversity activists has been litigation. It is important to clarify what is meant by litigation in this context. In

American society, litigation often has negative connotations because it is associated with one party suing another party over money. However, biodiversity protection litigation rarely seeks monetary damage awards or fines. Instead, litigation has been used by environmental groups as a form of law enforcement, appealing to the courts to compel a government agency to abide by and implement existing environmental laws when that agency has failed to do so. Key environmental laws such as the Endangered Species Act included citizen enforcement provisions specifically to enable the public to file lawsuits that would ensure that these laws were implemented. Although both the nationals and the grassroots groups used litigation to enforce biodiversity protection laws, the grassroots groups applied it much more extensively in cases that the national organizations avoided as too politically controversial, even though the legal claims were supported by scientific data. As Robin Silver of the Center for Biological Diversity summarized, "The use of science in a very aggressive fashion as the basis for litigation—no matter what the political risk—is what made us different." I examine how grassroots biodiversity groups' extensive use of litigation became central to their influence on federal environmental policies.

Organization

Large advocacy organizations are often the most visible manifestation of a social movement. However, some social movement researchers have cautioned that these large, bureaucratic organizations can actually constrain the movement from making social change.[17] While national environmental organizations such as the Sierra Club and The Wilderness Society received widespread publicity, many of the most notable victories in biodiversity protection in the 1990s and 2000s came from smaller groups. In this book, I consider the various organizational forms used by biodiversity activists—including large organizations, small groups, formal coalitions, and informal networks—and explore how these diffferent forms shaped their effectiveness.

Funding

Funding provides the resources for social movement organizations to carry out their activism. There were two primary funding sources used by the environmental groups in this book. Direct mail fundraising was a key source of support for the nationals, but the high costs of direct mail made this approach impractical for smaller groups. Instead, most grassroots biodiversity groups relied primarily on grants from philanthropic foundations. While these grants enabled the grassroots groups to do their work, some social movement scholars and activists have noted that foundation funding can also constrain advocacy.[18] Therefore, I assess how funding considerations influenced biodiversity activism.

Movement Culture

The choices that the members of a social movement organization make about how to approach their activism do not simply happen spontaneously. They are shaped and sustained by the larger community and the sets of ideas that encompass their organization (i.e., the movement culture). The actions of the biodiversity groups must therefore be considered within the context of their movement culture. I focus particularly on the role of the radical movement culture of Earth First! and how it fostered the outsider strategy of the grassroots groups.[19]

Political Conditions

The actions of the biodiversity groups were also shaped by the larger political conditions of the time. One key influence was the role of the varying political opportunities created by different presidencies. Throughout much of the 1990s, the White House was occupied by a Democrat who was generally considered supportive of environmental protections. Were the successes of the grassroots biodiversity groups during the 1990s simply the reflection of a sympathetic administration? To answer this question, it is helpful to compare the accomplishments of these groups under different administrations. Therefore, the case studies in this book focus on the period from 1989 to 2004, during which there were two presidential terms under Republicans (George H. W. Bush, 1989–1992; George W. Bush's first term, 2001–2004); and two under a Democrat (Bill Clinton, 1993–2000).[20]

The year 1989 serves as a logical starting point because that was the year of the first large-scale court injunction shutting down logging throughout the Pacific Northwest in order to protect the northern spotted owl. This case showed that environmental laws could be used to create sweeping changes in federal biodiversity protection policies and provided a key inspiration for the grassroots groups.

Although it was not apparent at the time when I began my research, 2004 would also mark a turning point for the environmental movement.

In the years immediately following 2004, there was rapid growth in public concern over global warming, spurred by the devastation from Hurricane Katrina, dramatic evidence of the rapid melting of the polar ice caps, and Al Gore's documentary, *An Inconvenient Truth*. Through the late 1980s and 1990s, forest protection was a primary focus of biodiversity activism, as epitomized by the struggles over the spotted owl and its old-growth forest habitat. After 2004, biodiversity protection would increasingly be defined in relation to climate change. In this context, the polar bear—a species whose existence is jeopardized by the loss of its Arctic sea ice habitat due to rising temperatures—became a central icon for the dangers of global warming. And the grassroots biodiversity groups that had played a central role in the era of the spotted owl would once again take the lead in the era of the polar bear.

The Death of Environmentalism or the Rebirth of Environmentalism?

The study of the grassroots biodiversity groups has much to offer. It fills a notable gap in the recent accounts of American environmentalism. One cannot properly understand the developments in federal environmental policy over the past twenty years without considering the role played by the grassroots groups. Moreover, the experiences of these groups offer valuable lessons for the environmental movement as a whole that may help guide it to become even more effective in the future.

Ironically, at the time that I was conducting the research for this book, there was a flurry of media attention around an essay titled "The Death of Environmentalism"[21] by professional consultants Michael Shellenberger and Ted Nordhaus, along with a related speech by their colleague Adam Werbach. Perhaps if those authors looked only at the behavior of the national environmental organizations, one could see how they might have reached such a dire conclusion. However, the experiences of the grassroots groups tell a very different story, and the authors were rightfully criticized for omitting the accomplishments of these groups.[22]

When reexamined in light of such contributions, we instead see an increase in the effectiveness of the environmental movement in recent decades due in large part to the new grassroots groups. These groups worked to overcome the false choice between being uncompromising advocates for the environment or being influential, a dilemma that had long hindered the movement. Thus, with the rise of the grassroots biodiversity groups, the 1990s and 2000s marked, not the demise of environmentalism, but instead its rebirth in a new form that was simultaneously bold and effective.

2

ORIGINS OF THE GRASSROOTS
BIODIVERSITY GROUPS

To understand the rise of the new grassroots biodiversity groups starting in the late 1980s, we must first situate those groups in relation to the other modes of environmental activism at that time. In the mid-1980s, there were two main approaches to environmental advocacy regarding the protection of wildlands and wildlife (i.e., biodiversity). The first approach was represented by the well-known national environmental organizations in Washington, DC that met together regularly in the 1980s as the "Group of Ten." This association consisted of the Sierra Club, National Audubon Society, National Parks and Conservation Association, The Wilderness Society, Izaak Walton League, National Wildlife Federation, Environmental Defense Fund, Natural Resources Defense Council, Friends of the Earth, and the Environmental Policy Institute.[1] The second approach to biodiversity activism was embodied by Earth First!, the primary radical alternative to the national environmental organizations in the 1980s. However, beginning in the late 1980s, a third approach emerged in the form of new grassroots biodiversity groups. As we will see in subsequent chapters, these grassroots groups would have a big impact on biodiversity protection policies in the United States over the next twenty years.

In this book, I do not focus on presenting the rationales for preserving biodiversity, nor do I examine why these activists chose to endeavor to save spotted owls, polar bears, and other species of wildlife from human-induced extinction.[2] Instead, I take their biodiversity protection goal as a premise and pursue a social movement analysis, exploring how the activists sought to achieve this goal so that we can better understand what enabled the grassroots groups to be so influential. In this chapter, after providing a brief overview of the origins of the modern American environmental organizations involved in biodiversity protection, I examine the three approaches to biodiversity activism in terms of their strategy, tactics, organization, funding, movement culture, and political conditions (see table 2.1). I present each approach as what sociologists call an "ideal type"—a generalized description of defining characteristics—in order to illustrate how the new grassroots groups were distinctly different from previous forms of biodiversity activism.[3] I then explore the specific histories of individual organizations through the in-depth case studies in the subsequent three chapters.

Origins of the Modern Environmental Movement

I begin by outlining the origins of the groups that would grow to become the national environmental organizations that later constituted the Group of Ten. Six of these organizations formed in the late nineteenth and early twentieth centuries as part of what was described at the time as the conservation movement. They were organized around a variety of outdoor interests. The oldest and perhaps best known of these groups is the Sierra Club, which was founded in 1892 by noted nature writer John Muir and a group of mountaineers in San Francisco who were concerned with the protection of California's Sierra Nevada mountain range. Around the same time, birdwatchers across the country organized various state-level Audubon societies, which then united in 1905 to become the National Audubon Society. The National Parks and Conservation Association was started in 1919 to provide oversight and support for the newly created National Park Service. The Wilderness Society, formed in 1935, was concerned with the preservation of undeveloped wild places outside national parks, particularly those on national forests. The interests of hunters and anglers were represented by the Izaak Walton League, founded in 1922, and the National Wildlife Federation, which began in 1936. For the first half of the twentieth century, the membership rolls of these early conservation groups were generally quite small. For example, the Sierra Club had fewer than 3,000 members in the 1930s and fewer than 7,000 members in 1950.[4] Most of these groups had few, if any, full-time staff. Some, such as The Wilderness Society and the National Parks and Conservation Association, depended on the support of a handful of wealthy patrons. The political influence of these early groups was generally limited.

Table 2.1
Three paths to biodiversity protection in the U.S. environmental movement

	NATIONAL ENVIRONMENTAL ORGANIZATIONS	EARTH FIRST!	GRASSROOTS BIODIVERSITY GROUPS
Strategy	Insider strategy	Outsider strategy	Outsider strategy
Tactics	Conventionally prescribed forms of political participation, particularly lobbying. Application of litigation constrained by the insider strategy.	Direct action, including "monkey-wrenching."	Lawful tactics, particularly litigation, but applied more extensively without the political constraints of the insider strategy.
Organization	Large, bureaucratic organizations, centered around a professional staff.	No formal organization. No staff. No official members. Membership defined through participation.	Small formal organizations with radical staff, often coming from Earth First!, generally without professional credentials, working for very low pay.
Funding	Very large budgets. Membership donations obtained through direct mail solicitations are a primary funding source.	Scant resources. Ineligible for foundation grants. Some funding from sales of the *Earth First! Journal* and non-tax-deductible donations.	Limited funding, primarily coming from grantmaking foundations, particularly maverick new funders whose staff have roots in environmental groups that use an outsider strategy.
Movement Culture	Diminished movement culture. Members have little or no direct involvement with the organization, while staff are increasingly motivated by careerism.	Radical movement culture of music, art, and philosophy. Community built through national gatherings and open dialogue in the *Earth First! Journal*.	Influenced by the radical movement culture of Earth First!, but most groups do not develop significant community-building institutions of their own.
Political Conditions	Early development during a time when environmentalism is a nonpartisan issue. Insider organizations become increasingly tied to the Democratic Party amid the ascendancy of an antienvironmental wing of the Republican Party.	Develops out of disappointment with shortcomings of environmental protection during the Carter administration, followed by concerns over the threat posed by the antienvironmental wing of the Republican Party under Reagan.	The election of Clinton creates conditions under which grassroots groups are better able to distinguish themselves from the nationals, though those differences subsequently become more muted during the George W. Bush administration.

That situation began to change in the 1950s with the Sierra Club's Echo Park campaign. After the death of John Muir in 1915, the Club entered a long period during which it focused primarily on organizing hiking trips rather than on environmental advocacy. The Sierra Club then returned to an activist role in national park protection when David Brower became the Club's first executive director in 1952. Under Brower's leadership, the Club began an aggressive campaign to stop the construction of the Echo Park dam, which threatened to flood part of the Dinosaur National Monument in Utah. Brower mobilized public opposition to the dam using innovative techniques, including the production of a large-format book filled with color photographs of the scenic places that were threatened to be submerged by the dam.

The Club's campaign successfully organized enough public opposition to stop the Echo Park dam in 1956. This was the first time that an environmental group had blocked a major federal project. As a concession, however, the Sierra Club's board of directors agreed not to oppose a dam project in Glen Canyon in Arizona. Glen Canyon was not in a national park, but Brower urged the board to fight that dam as well. He was overruled. Board members argued that a stronger position was politically unfeasible and that "the compromise had to be kept if the Club was to preserve its credibility."[5] The subsequent flooding of this spectacular canyon by the dam led Brower to question the role of compromise in environmental advocacy and embrace a more confrontational approach. The Echo Park campaign thus had a dual legacy. On the one hand, it marked the beginning of modern environmentalism as a significant political force. On the other hand, it also foreshadowed the problematic compromises that often accompanied this political activism.

Echo Park provided inspiration for the Sierra Club and other conservation groups to undertake new political campaigns. For example, The Wilderness Society spearheaded a drive for legislation to create designated wilderness areas on federal lands that would be protected from logging, road building, and other forms of development. The first version of this bill was introduced two months after the Echo Park dam was stopped, but the Wilderness Act ultimately did not become law until 1964. During this time, conservation groups worked to mobilize public interest in the bill. In 1962, Congress got more mail about wilderness protection than any other issue.[6] However, passage of the bill was a long, grueling process that involved many compromises to appease opponents. As historian Paul

Hirt noted, "The Wilderness Act has been touted as a victory for preservationists and a defeat for advocates of total use, but this assessment can be overstated . . . the Act was loaded with compromises. In fact, in the context of national forest management, the victors initially gained little and the defeated lost even less."[7] One notable compromise was that while the bill's proponents had originally intended that additional wilderness areas would be created by executive orders from the president, in the final language such new areas required legislative approval by Congress. This change meant that the energies of many of the largest conservation organizations would increasingly be channeled into legislative campaigns.

As conservation groups such as the Sierra Club and The Wilderness Society pursued a more outspoken approach to environmental advocacy, their membership numbers swelled in the 1960s. At the same time, there was also increased public awareness of pollution, propelled by the popularity of Rachael Carson's bestselling 1962 book about pesticides, *Silent Spring*, and culminating at the end of the decade with extensive publicity around a large oil spill on California's coast and the Cuyahoga River in Ohio catching fire due to the many chemicals dumped into it. All of these developments contributed to a burgeoning public interest in environmentalism. The breadth of support for environmentalism was made evident by the first Earth Day celebration in 1970. The event sought to apply the effective "teach-in" tactic of the antiwar movement to environmental issues. Earth Day would be a national teach-in with ecological education events held simultaneously around the United States on April 22, 1970, a date chosen to coincide with John Muir's birthday. Ironically, many of the largest conservation groups, including the Sierra Club, National Wildlife Federation, and National Audubon Society did little or nothing to organize the first Earth Day (though The Wilderness Society provided some financial support).[8] Instead, the success of Earth Day was the result of grassroots initiative. As Senator Gaylord Nelson, who first proposed the event, recounted, "Once I announced the teach-in, it began to be carried by its own momentum. If we had actually been responsible for making the event happen, it might have taken several years and millions of dollars to pull it off. In the end, Earth Day became its own event."[9] Earth Day 1970 far exceeded the organizers expectations. More than 20 million people participated in Earth Day events.[10] It marked the arrival of environmentalism as a major force in American society.

Of the organizations that would constitute the Group of Ten in the 1980s, the remaining four—Environmental Defense Fund, Natural Resources Defense Council, Friends of the Earth, and Environmental Policy Institute—formed around the time of the first Earth Day. The development of these groups was strongly shaped by the new funding opportunities from philanthropic foundations that took an interest in the growing environmental movement. The most significant influence came from the Ford Foundation, which was largely responsible for the emergence of the Environmental Defense Fund and Natural Resources Defense Council, as well as a similar organization, the Sierra Club Legal Defense Fund.

The Environmental Defense Fund grew out of the efforts of a small group of birdwatchers and scientists to stop the spraying of the pesticide DDT on Long Island because of the detrimental impact on birds. They were joined by attorney Victor Yannacone, whose hallmark phrase—"Sue the bastards!"—epitomized his aggressive approach to using litigation in support of environmental protection. Yannacone won an injunction against local DDT spraying, which catalyzed opposition to DDT nationally and ultimately led to a ban on DDT in the United States. Buoyed by this success, he tried to develop a litigation arm within the National Audubon Society. However, Audubon's board, which included directors with connections to chemical companies, thwarted this move.[11] Instead, Yannacone received start-up funding from the Ford Foundation to create the Environmental Defense Fund in 1967, but his confrontational style soon raised concerns with the foundation and the Environmental Defense Fund's board of directors. In 1970, he was fired. Soon after, the Ford Foundation gave the organization a large grant, but the money came with one key condition—the Environmental Defense Fund would have to submit to a litigation oversight panel created by the Ford Foundation, known informally as the "Gurus."

The Gurus also exercised oversight over two other new organizations funded by the Ford Foundation—the Sierra Club Legal Defense Fund and the Natural Resources Defense Council. The Sierra Club Legal Defense Fund formed as the litigation arm of the Sierra Club (though it would also work with other environmental groups) with Ford Foundation seed money in 1971. The origin of the Natural Resources Defense Council provides an even clearer example of the level of influence that Ford exercised over the new groups. The Natural Resources Defense Council was an arranged marriage of two proposals received by the Ford Foundation. The first proposal came from a group of recent graduates of Yale Law School, led by Gus Speth, who were eager to prepare litigation in support of environmental protection campaigns. The second proposal came from a pair of Wall Street attorneys who, while fighting the proposed Storm King power plant near their homes, had established an important legal precedent that allowed environmental groups to have standing to sue environmentally harmful projects on behalf of the public interest. Their victory in that case led them to form an environmental litigation organization called the Natural Resources Defense Council. In 1970, Ford agreed to fund this organization, but insisted that the Yale graduates from the other proposal be hired as its staff. In this context, the organization's founders, along with the oversight provided by Ford's Gurus, could provide a moderating influence on the young lawyers from Yale. The hands-on role of Ford was characteristic of the new circle of foundations that expected to play a more active part in overseeing and shaping their grantees in the environmental movement.

The two other organizations that became part of the Group of Ten developed from two successive organizational splits. In 1969, David Brower was forced out as executive director of the Sierra Club by its board of directors, who were troubled by Brower's confrontational approach to environmental advocacy and his management of the organization. That year, Brower started Friends of the Earth, which took on a more international focus than most other environmental groups. In 1972, part of Friends of the Earth's Washington, DC staff split off from the organization over their frustration with Brower's wariness of letting the group become too involved in political deal-making. The breakaway faction formed the Environmental Policy Institute, although they later remerged with Friends of the Earth in 1989 after Brower had left the organization.

National Environmental Organizations

The success of Earth Day in 1970 led to two notable outcomes that transformed this diverse set of groups into the national environmental organizations as we know them today. First, there was a surge in the membership of the largest environmental groups, where the total number of members increased by 38 percent between 1969 and 1972.[12] (I discuss this point at greater length later.) The second major development following Earth Day was an outpouring of federal legislation on environmental protection. Twenty-three major environmental laws were enacted during the 1970s, including the National Environmental Policy Act, the Marine Mammal Protection Act, the Endangered Species Act, and the National Forest Management Act. This environmental legislation is often presented as one of the foremost accomplishments of the national environmental organizations and a reflection of their political clout following Earth Day. However, a closer examination of the legislative history reveals that the dominant national organizations played minor (and at times even counterproductive) roles in the passage of those laws (see appendix).

While the national environmental organizations did not play a central role in passing those laws, the environmental laws of the 1970s created a federal environmental policy apparatus that drew in the national environmental organizations and led to a focus on activity in Washington, DC. As Brock Evans—a longtime DC-based staff member with the Sierra Club and later the National Audubon Society—explained,

> We must be there, because it is Congress that ultimately decides which areas should be logged and which shall remain wild; it is the EPA that promulgates the vital air and water pollution regulations; and it is the President himself and his aides who, by a phone call, can often determine the fate of a bill in Congress or a policy in a bureaucracy.[13]

In this context, the national environmental organizations embraced an "insider" strategy to try to achieve environmental protection. The insider strategy is commonly used by most interest groups in Washington, DC that seek to shape government policies, so it is not surprising that the large environmental organizations followed this path. An organization pursuing an insider strategy depends on building relationships with politicians and other decision makers in order to influence political deal-making. Privileged access to decision makers is seen as the primary measure of the insider's power. Through that access, the national environmental organizations hoped to gain a seat at the negotiating table so as to be able to shape the newly created federal environmental policy apparatus.

The intended result was incremental progress toward environmental protection achieved through political deal-making and compromise. (Indeed, one of the primary bargaining chips that the national organizations could offer in such negotiations was their ability to persuade their members to accept these compromises.) However, the role of compromise in this strategy attracted some critics, including former Sierra Club Executive Director David Brower:

> The problem with too many environmentalists today is that they are trying to write the compromise instead of letting those we pay to compromise [i.e., politicians] do it. They think they get power by taking people to lunch, or being taken to lunch, when in reality they are just being taken. They don't seem to learn, as I did at Glen Canyon, that whenever they compromise they lose.[14]

Despite such criticisms, a commitment to an insider strategy became the defining characteristic of the national environmental organizations.

The insider strategy relied on conventional forms of political participation. The primary tactic was lobbying. The role of lobbying in the environmental movement changed quite dramatically in the 1970s. Prior to that time, outreach to elected officials was generally done on an irregular basis by volunteers. In 1969, there were only two full-time environmental lobbyists in Washington.[15] Environmental groups during that time were reluctant to engage in too much lobbying for fear that they would lose their tax-exempt status, which in turn would mean the loss of funding from grantmaking foundations and large donors. At the beginning of the 1970s, tax-exempt groups spent less than 5 percent of their budgets on lobbying, but changes in the Tax Reform Act of 1976 subsequently allowed them to spend up to 20 percent. The change in the tax laws was a watershed moment for the national environmental organizations. They responded with a sharp increase in lobbying activity, and groups such as Audubon immediately opened lobbying offices in Washington. By 1985, the nationals employed eighty-eight lobbyists.[16]

Even the newer national organizations that initially focused on litigation hired lobbyists. At the same time, some of the older conservation organizations joined in filing environmental lawsuits. Because of the new environmental laws, litigation became a valuable tool for the environmental movement. However, as the insider strategy developed, the use of litigation by these organizations was limited by political considerations. For example, Dave Foreman—a Wilderness Society lobbyist in the late 1970s who later cofounded Earth First!—recalled how political considerations constrained the national environmental organizations' response to an unfavorable decision by the Forest Service on the protection of road-less national forest lands (known as RARE II):

> We were worried that some local wilderness group might go off the reservation and sue the Forest Service over the clearly inadequate environmental impact statement for RARE II. We didn't want a law-suit because we knew we could win and were afraid of the political consequences of such a victory. We might make some powerful sen-ators and representatives angry.[17]

While the legacy of Earth Day–era legislation shaped the strategy and tactics of the national environmental organizations, the surge in public sup-port for those groups following Earth Day transformed their funding, orga-nizational structure, and movement culture. The nationals increasingly sought new members through the mass mailing of funding solicitations to the public, known as "direct mail." By the early 1990s, the largest national organizations received about 65 to 70 percent of their budgets through direct mail appeals.[18] Direct mail led to tremendous organizational growth for the nationals. At the same time, the growing dependence on direct mail-based funding brought with it new implications for the environmental movement. As one direct mail copy writer summarized, "direct mail became the move-ment's heroin."[19] The mass mailings brought in members who lacked a strong stake in the organization. As a result, environmental groups relying on direct mail faced a 30 to 40 percent membership attrition rate each year, leading to a constant struggle to maintain membership levels.[20] This dependence on continuously attracting new members in turn affected which issues the organizations chose to focus on.[21] If an environmental issue was seen as not likely to produce a sufficient return from direct mail fundrais-ing, it was far less likely to be adopted as a campaign. Likewise, when an issue generated a strong direct mail response, many groups gravitated toward it. While the various national environmental organizations began with fairly distinct niches at their inception, those differences became muted when they all shifted to whatever issue was most likely to generate a favorable return at the moment. For example, in 1991, six large environmental organ-izations were simultaneously focused on the Arctic National Wildlife Refuge.[22] As a result, direct mail contributed to homogenization of the national environmental organizations.

Direct mail fundraising also contributed to homogenization by favoring large, established environmental organizations over newer groups. Direct

mail operates most effectively on a mass scale, depending on sending solicitations to thousands or even millions of people at a time. A 2-percent
response rate is considered successful for direct mail, so only a very large
mailing will recoup the preparation costs. An organization therefore needs
to have extensive resources to cover the large up-front costs. New groups
without a large membership base also need to purchase expensive mailing
lists in order to reach potential members. And overall, larger and older groups
have a strong advantage in direct mail fundraising because of their more
established name recognition. As a result, while direct mail contributed to
the rapid expansion of the large established environmental organizations,
it simultaneously created barriers-to-entry for newer and smaller groups.

As the national environmental organizations grew larger, they became
more bureaucratic. The activities of these organizations became focused
on an increasingly professionalized staff. The insider strategy required a
select group of specialists to cultivate privileged relationships with policy
makers, so these organizations became more staff-centered. In addition to
lobbyists, they employed scientists, lawyers, and other professionals who
were seen as offering the organizations greater credibility in the eyes of
policy makers. Staff members were also more likely to be hired based on
their experience in political or business circles rather than within the environmental movement. For the staff, work in environmental advocacy
became a career. They often traveled through a revolving door of jobs with
various environmental organizations, other types of interest groups, and
positions in government. This shift in staffing was particularly evident in
the highest echelons of these organizations. By the 1980s, all of the primary national environment organizations had replaced their previous leaders with new directors conforming more closely to the model of the
corporate CEO.[23]

Meanwhile, the role of the membership within these organizations
receded. In the earlier days of the conservation movement, volunteers
had played a primary role in these groups. Historian Stephen Fox concluded that "the radical amateurs comprised the heart and soul of American conservation."[24] While some older groups like the Sierra Club and
Audubon continued to organize outings for hikers and birdwatchers, by
the 1980s most environmental group members played no notable role in
the organizations to which they belonged. Participation for these members was generally quite minimal, consisting of a check written once a
year in exchange for receiving a glossy magazine touting the organization's

accomplishments, a tote bag with the organization's name on it, and the occasional preprinted advocacy postcard that needed only to be signed and returned to the organization. There were no regular opportunities for face-to-face contact with other members in most of these groups and members had little or no say in the decisions of their organizations.[25]

These organizational changes took place amid shifting political conditions that altered the nationals' relationships to the major political parties. In the 1960s and early 1970s, environmentalism was not seen as a partisan issue. For example, the Wilderness Act had benefited from the support of many eastern Republicans as well as Democrats, while its chief opponent had been a western Democrat. Social movements are most likely to have political success when both major political parties are actively competing for the votes of the movement's members.[26] However, these conditions began to shift for the environmental movement in the 1970s and early 1980s. Although Republican president Richard Nixon signed most of the major environmental laws of the 1970s, many environmentalists were particularly enthusiastic when Democrat Jimmy Carter was elected president in 1976. For the first time, substantial numbers of staff from the national environmental organizations were appointed to positions in a presidential administration. However, this link to the Carter administration would be short lived. Carter was defeated in 1980 by Republican Ronald Reagan.

The election of Reagan marked the ascendancy of a wing of the Republican Party that was actively hostile to the environmental movement. This crisis provided the impetus for a cadre of foundations to broker the creation of the Group of Ten in order to foster a more coordinated response by the nationals. These organizations then took on a high-profile campaign against the Reagan administration, particularly the flamboyantly antienvironmental secretary of the interior James Watt. Watt eventually resigned after making too many ill-considered public statements that embarrassed the administration. Meanwhile, the nationals were able to turn the situation into a fundraising bonanza. Many of the larger organizations experienced rapid growth in membership between 1980 and 1983: The Wilderness Society, 144 percent; Sierra Club, 90 percent; and Defenders of Wildlife, 40 percent.[27] At the same time, the insider environmental organizations in Washington solidified their connections with the Democratic Party, which still held a majority in Congress.

In the short term, this approach of criticizing the Reagan administration and supporting Democrats in Congress worked fairly well for the nationals in terms of boosting their funding and political access. However, it also meant that the environmental movement was increasingly tied to the Democratic Party. Under these conditions, Republicans were less likely to court environmental voters, and in the absence of electoral competition, Democratic politicians could more easily take their environmentalist constituents for granted. The insider strategy became less effective in this political context, and those problems would only be magnified when the Democrats lost control of Congress in the 1990s.

In the meantime, the national environmental groups had achieved tremendous organizational success. In the twenty years since the first Earth Day, their membership numbers had grown manyfold. For example, the Sierra Club had increased from 83,000 members in 1969 to 493,000 members in 1989. The nationals held substantial resources, with the Group of Ten possessing a combined budget of more than $200 million by 1990.[28] They were treated as the official voice of the environmental movement by the media. They had large professional staffs with access to prominent policy makers. And they had established themselves as a well-recognized lobbying force in Washington. However, their dependence on an insider strategy and the resulting political compromises were attracting growing criticism from a more radical wing of the environmental movement, as embodied by Earth First!

Earth First!

Earth First! began in 1980 as a critique of the national environmental organizations. All but one of the five founders of Earth First! had worked on wilderness issues for national organizations such as The Wilderness Society, Sierra Club, and Friends of the Earth in the 1970s.[29] There they had directly experienced the increasing professionalization of those organizations, combined with the growth of the insider strategy. Their frustration with these developments led them to form Earth First!

In particular, a key source of frustration for wilderness advocates was the outcome of the Forest Service's second Roadless Area Review and Evaluation process (RARE II) in the late 1970s. RARE II was conceived by M. Rupert Cutler, a Carter administration official who had previously been an assistant executive director of The Wilderness Society. Under

RARE II, the Forest Service was in the process of deciding which of its 80 million acres of the remaining undeveloped lands without roads would be eligible to become official wilderness, and which areas would be opened up to logging and other forms of commercial extraction. The overall strategy of the national environmental organizations was to present a united front on which areas they were requesting be protected by the Forest Service. As sociologist Rik Scarce notes, "that entailed accepting the lowest common denominator, the weakest positions of the bunch, to keep everyone together."[30] The nationals pressured smaller regional wilderness groups to reduce the amount of wilderness they proposed for their region so that the final proposal would appear more "reasonable."[31]

The environmentalists proposed that about half of the eligible roadless areas become wilderness.[32] Despite the success of the nationals in moderating the proposal, their approach produced disappointing results. In 1979, the Forest Service chose to only recommend 15 million acres for wilderness—less than a quarter of the potential area, and much of that was alpine "rock and ice" areas with no possible commercial use. At the same time, the Forest Service sought to immediately open up 36 million acres to road building, logging, and other development.[33]

The deep disappointment among environmental activists over the outcome of RARE II stoked misgivings about the insider strategy of the national environmental organization. Howie Wolke, one of the cofounders of Earth First!, was a regional organizer for Friends of the Earth during RARE II. As he recounted,

> We played the game, we played by the rules. We were moderate, reasonable, professional. We had data, statistics, maps, graphs. And we got fucked. That when I started thinking, "Something's missing, here. Something isn't working." That's what led to Earth First! more than anything else.[34]

The outcome of RARE II left Earth First!'s founders frustrated not only by the nationals' strategy, but also by the decisions made by the Carter administration despite having former staff of the environmental groups working within that administration. It raised doubts about the prospects for sufficient environmental protection even under a seemingly sympathetic administration. Disappointment with the Carter administration created the context for the emergence of Earth First!, but the election of Ronald Reagan in 1980 provided a key impetus for Earth First!'s growth. If the national environmental organizations had not been effective enough during the more favorable Carter years, their insider strategy appeared unlikely to protect the environment in the face of a hostile administration.

By 1980, the five cofounders of Earth First! had grown tired of trying to appear reasonable to politicians. After a trip together in the wilderness, they developed a far more sweeping proposal for large-scale wildlands protection than the truncated proposals of the nationals. At its inception, Earth First! was created as a means to put forward that vision. Rather than being a formal organization per se, it began as an attitude and a shared identity, embodied in its slogan, "No compromise in defense of mother earth." That slogan exemplified Earth First!'s rejection of the insider strategy. Earth First! was not concerned with having access to politicians. It was not interested in getting a seat at the table in order to bargain and offer concessions. It eschewed compromise as a means for environmental protection. Above all, Earth First! did not avoid controversy. Indeed, being controversial was seen as a way to draw more attention to environmental issues. This approach positioned Earth First! as an outsider to the Washington policy-making process. The challenge that Earth First! faced then was to find a means for protecting biodiversity outside of conventional forms of political participation. This "outsider" strategy was largely unexplored terrain for modern environmentalists, and Earth First! was by necessity experimental.

Much of Earth First!'s experimentation took place at the level of tactics. Since Earth First! worked outside the political system, it was not precluded from pursuing illegal tactics. Rather than trying to appeal to politicians to protect wildlife and wildlands, Earth First!ers took action to directly stop environmentally destructive activities. Earth First!'s direct action had two notable forms. One was civil disobedience, in which Earth First!ers used their own bodies to actively obstruct logging and other activities that harmed wildlife—for example, standing between a logger's chainsaw and the tree, chaining themselves to a bulldozer, or sitting high up in trees that were scheduled to be cut down. These actions were not simply symbolic. In these situations, the environmentally destructive activity would have to be halted or else the Earth First!ers could be injured or even killed. While it was rare that Earth First!'s actions alone would save a forest, a sustained direct action campaign could sometimes delay the trees from being cut long enough for other environmental groups to get a court injunction to stop the logging project. These risky actions also received media coverage, which helped draw public attention to biodiversity issues such as the logging of old-growth forests.

However, the visibility of this type of direct action was initially over-shadowed by a second tactic known as "monkeywrenching," which entailed stopping environmentally destructive activities by damaging or destroy-ing the equipment involved in the activity. Such activities were done clan-destinely to avoid arrest. The goal of monkeywrenching activities such as putting sand in a bulldozer's gas tank or pulling up survey stakes was to raise the costs of environmentally destructive activities to a point where the companies involved would no longer find those activities profitable. In practice, the actual economic effect from monkeywrenching was unclear, and monkeywrenching on its own saved few, if any, places.[35]

Nonetheless, monkeywrenching became a quintessential part of Earth First!'s public image in the 1980s. But as sociologist Timothy Ingalsbee explained, it was a problematic identity: "Monkeywrenching was a core iden-tity construct that partially defined the movement's radical activism, and brought public and media attention to it far in excess of the movement's size and strength or the actual incidents of sabotage . . . few [Earth First!] activists and almost no organizers actually do any monkeywrenching."[36] The media focused on monkeywrenching because it defied the conventional atti-tudes toward private property and had an element of danger. The publicity swelled the ranks of Earth First!'s participants and supporters. However, the emphasis on monkeywrenching gave a distorted view of Earth First!'s activ-ities, and it also attracted greater scrutiny by the FBI. As public relations consultant Herb Gunther summarized, "Early on the monkeywrenching gave them a boost because it got them attention that they would have never got-ten otherwise. But then it started working against them."[37]

Earth First!'s tactics shaped its organization. The use of illegal tactics precluded it from developing a formal organizational structure because such an entity would have been quite vulnerable to a crackdown from the authorities. At the same time, Earth First!'s informal structure was a direct response to the problems its founders had experienced with the profes-sionalization and bureaucratization of the national organizations. Indeed, Earth First! was the antithesis of the nationals. It lacked any formal organ-ization and did not seek nonprofit status. It also had no official members. Instead, being an Earth First!er was defined through direct participation. An Earth First!er was someone who identified with Earth First!'s values and took part in its activities.

While Earth First! had no formal staff or leadership, like other unstructured groups, it developed informal leaders and media celebrities.[38] Dave Foreman—one of Earth First!'s cofounders and most charismatic speakers—became the public face of Earth First! during the 1980s. Foreman emphasized Earth First!'s association with monkeywrenching and promoted an interpretation of Earth First!'s philosophy of deep ecology (discussed later) that was particularly antagonistic to social justice considerations. He represented one perspective within the diverse community of Earth First!ers, but the extensive media attention on Foreman created the misperception that his positions were indicative of Earth First! as a whole.

At the same time, the lack of formal structure provided a considerable freedom for Earth First!ers to experiment with unconventional ideas and tactics away from the constraints of a traditional environmental organization. There were three components of Earth First! that held this loose network of activists together in a shared identity: an annual gathering called the Round River Rendezvous; a national publication, the *Earth First! Journal*; and a unifying philosophy known as deep ecology. These three elements were integral to Earth First!'s movement culture.

The concept of movement culture—the community-building facets of a social movement that nurture the shared values and ideas that motivate social change activists—had faded in the environmental movement in the 1970s as the national environmental organizations became more professionalized and the role of the organizations' members was largely reduced to paying dues. In contrast to the national organizations, Earth First! developed a vibrant movement culture in 1980s. It included a troupe of Earth First! musicians whose feisty and often humorous songs celebrated Earth First!'s rebellious stance and lampooned its opponents. The musicians visited regional Earth First! groups as part of traveling road shows that also featured Earth First! speakers. Then once a year, Earth First!ers would come together for a national gathering in the wilderness known as the Round River Rendezvous. Here Earth First!'s rowdy campfire culture gained full expression in a mixture of music, speeches, satirical performances, trainings, conversation, and celebration. By tradition, the annual rendezvous concluded with the participants joining together to engage in civil disobedience against an environmentally harmful project near the location of the gathering.

During the rest of the year, Earth First!ers remained connected to each other through the *Earth First! Journal*. It featured works by Earth First!'s artists, poets, and movement intellectuals, as well as reports on direct action campaigns and techniques for monkeywrenching. It contrasted sharply with the national environmental organizations' publications. While the latter produced glossy magazines filled with color pictures of scenic landscapes, the journal was a newspaper-format "rag" with black-and-white photos of protests. Whereas the nationals' magazines served primarily as a vehicle to tout their accomplishments to their members, the *Earth First! Journal* offered an unusually free space within the environmental movement for dialogue and debate on new ideas, no matter how controversial.

One of the ideas frequently discussed in the journal was deep ecology, a concept introduced by philosopher Arne Naess in the 1970s and subsequently popularized by Earth First! activists in the 1980s.[39] A central premise of deep ecology that appealed to Earth First!ers was the idea of biocentrism, the belief that all species of life have inherent worth and an intrinsic right to exist. For example, from that perspective, if the logging of the old-growth forests where spotted owls live will cause that species of owl to go extinct, then that logging should be stopped even if the timber produced by that logging is commercially valuable and the owl itself has no direct monetary value. Biocentrism is contrasted with anthropocentrism, the idea that other species matter only to the extent that they are useful as a resource for humans. Earth First!ers' interest in deep ecology was reinforced by concurrent scientific research on the importance of maintaining a wide variety of species for the well-being of ecosystems.[40] Deep ecology became a unifying philosophy for Earth First!. It offered an intellectual framework for advocating for wildlife and wild places without having to justify those goals in terms of immediate benefits to people. Within a political system grounded in interest group politics, that stance was liberating for biodiversity activists. Identifying as a deep ecologist signaled one's unconstrained commitment to the protection of other species and provided a collective identity among Earth First!ers.

Ultimately, the most influential aspect of Earth First!'s movement culture was the way that it nurtured an outsider strategy to social change. Earth First! opened an alternative space for environmental activism without the constraints of the moderate national organizations. Within this space, the movement culture of Earth First! encouraged questioning of the dominant forms of environmental advocacy, and it fostered experimentation with alternate approaches.

While Earth First! had a rich movement culture, it was not rich in funding sources. Because of its illegal tactics and because it was not a non-profit organization, it could not receive support from grantmaking foundations. Furthermore, on principle, its founders had rejected the direct mail fundraising that sustained the nationals. Instead, a primary source of funding for Earth First! came from Earth First! Journal subscriptions, along with sales of Earth First! t-shirts (featuring the group's iconic raised fist logo), bumperstickers (e.g., "Nature Bats Last"), and other trinkets.

The journal was run as a for-profit business, and by 1988, it had a budget of more than $200,000.[41] The income provided funding for some Earth First! actions and modest stipends for a few organizers. For example, in 1985, Dave Foreman was receiving $250 per month.[42] Earth First!ers also established a tax-deductible entity initially called the Earth First! Foundation—later renamed the Fund for Wild Nature—that provided support for research, advocacy, and educational work by Earth First! activists. By 1988, it distributed $50,000 per year.[43] Although this foundation was much smaller than the dominant environmental grantmakers, it provided seed money for projects that grew into some of the grassroots biodiversity groups in the 1990s.

Despite these creative fundraising strategies, without a dues-paying membership or foundation grants, Earth First! was in no position to hire full-time staff. As such, Earth First! embodied the resurgence of the "radical amateurs" within an increasingly professionalized environmental movement. At the same time, this arrangement meant that activists had to look elsewhere for a source of income to sustain themselves, leaving Earth First! for their spare time.

In its first decade of existence, Earth First! grew rapidly. What began as a handful of frustrated environmentalists dreaming up grandiose wilderness proposals around a campfire had become an entity that garnered national (and even international) attention. By 1990, there were estimated to be 15,000 Earth First!ers.[44] At the same time, Earth First! was also facing difficult challenges. In 1989, five Earth First!ers, including Dave Foreman, were arrested for conspiring to cut down power lines in Arizona. Foreman was not directly involved in the action, which was organized by an FBI infiltrator. Nonetheless, Foreman was charged with providing funding for it. He avoided jail in a 1991 plea bargain in which he agreed to stop advocating monkeywrenching. By then he had already publicly departed from Earth First!.

Foreman's departure marked a significant shift within Earth First!. In the 1990s, Earth First! moved away from monkeywrenching and reemphasized civil disobedience as a way to increase participation and build broader coalitions. This new phase of Earth First! played a prominent role in some notable biodiversity protection campaigns of the 1990s, such as the Headwaters Forest campaign (see chapter 3). However, Earth First!'s most important influence came through the activists who had participated in its community and then went on to form grassroots biodiversity groups.

Grassroots Biodiversity Groups

Environmental activists in the late 1980s faced a choice between two paths for working on the protection of biodiversity, but both paths had notable shortcomings. One path, represented by the national organizations, provided political access but was hindered by compromise and constraint. The second path, embodied by Earth First!, offered an unconstrained approach to biodiversity advocacy, but it was not particularly influential. However, some activists found a way to transcend this dilemma. They created a third path by forming grassroots biodiversity protection groups. Unlike the national organizations, the small, new groups did not usually have offices in Washington, DC, and were not invested in insider politics. Unlike Earth First!, they relied on legal tactics, particularly litigation. Some examples of grassroots biodiversity groups were the Biodiversity Legal Foundation, Center for Biological Diversity, Forest Conservation Council, Forest Guardians, Heartwood, John Muir Project, Native Forest Network, Southern Appalachian Biodiversity Project, and Wild Alabama.

For decades there had been small environmental associations that generally limited themselves to local issues, but starting in the late 1980s there was a proliferation of grassroots biodiversity groups that were distinctly different from previous small organizations. The new entities actively engaged in shaping regional and national environmental policies, and they took a notably more confrontational approach to environmental protection. This approach reflected their underlying outsider strategy. Many of the founders and early staff members of these groups had participated in Earth First! or drew inspiration from its unconstrained environmental advocacy. These activists had been deeply influenced by Earth First!'s movement culture and its critique of moderate environmentalism. From this, they embraced an outsider strategy for social change which did not depend on appealing to politicians and avoiding controversy.[45]

While they embraced Earth First!'s outsider strategy, the grassroots biodiversity groups were distinctly different from Earth First! in terms of their tactics and organization. As Todd Schulke—an Earth First! activist who later cofounded the Center for Biological Diversity—recalled,

> Here were all these people who were committed to the point that we were taking great risks. And we realized fairly quickly how limited an influence we had standing on the front lines doing direct action. Not that it was completely irrelevant, but as an end it didn't get us nearly what we expected it would. So I think a lot of people took that passion and moved it into different approaches. The philosophy stayed the same, but the approach changed dramatically.

While Earth First! in the 1980s largely defined itself through monkeywrenching and other forms of direct action, the grassroots biodiversity activists explored alternate tactics. In particular, they found administrative appeals and lawsuits to be powerful tools to compel the increased enforcement of existing environmental laws. Activists discovered a variety of legal techniques for challenging logging and other activities that harmed biodiversity, including applications of the National Environmental Policy Act, the Endangered Species Act, and the National Forest Management Act, as well as the Marine Mammal Protection Act for ocean issues (see appendix). Litigation became the defining tactic of the grassroots biodiversity groups.

At first glance, such law-abiding tactics might appear to be a rejection of Earth First!'s radicalism, but closer examination reveals that was not the case. For example, a successful lawsuit could stop an entire logging project. In this regard, litigation had a similar effect to the Earth First!ers who sat in trees. Both tactics directly blocked environmentally harmful activities, and both were rooted in a strategy that did not rely on trying to convince politicians to act on their behalf. However, backed by the authority of the courts, litigation could have an even larger and more lasting impact than direct action.

The potential power of this tactic became readily apparent in 1989 when federal judge William Dwyer issued his first injunction against logging in national forests in the Pacific Northwest that were home to the imperiled northern spotted owl (see chapter 4). Dwyer's injunctions would ultimately stop most old-growth logging in the Pacific Northwest until 1994. The success of the spotted owl litigation was a key inspiration for the proliferation of grassroots biodiversity groups in the years that followed.

Litigation was an appealing tactic for the grassroots groups because it did not necessarily require extensive resources. Grassroots activists generally did not have sufficient funding to hire attorneys, but that was not necessarily a barrier to litigating. Environmental laws such as the Endangered Species Act included citizen-suit provisions requiring the federal government to reimburse the fees of lawyers who successfully sued it for failing to enforce those laws. As a result, some attorneys were willing to take on biodiversity protection cases at no charge to the activists with a reasonable hope that they would ultimately recoup their fees.

The grassroots groups' aggressive approach to litigation was markedly different from that of the moderate national environmental organizations. While the nationals sometimes filed lawsuits, they were often reluctant to pursue cases that might stir up controversy and upset influential politicians. By contrast, because the grassroots groups did not depend on insider access to politicians to achieve their goals, these groups did not need to avoid controversy. Therefore, there were abundant opportunities for the grassroots groups to take the lead in initiating litigation to protect forests or wildlife in situations that the national organizations either overlooked or avoided as being too controversial.

As grassroots activists started doing litigation to protect forests and endangered species, they created new organizations as vehicles for these lawsuits. The new groups were initially very small with few members other than the activists directly involved in the group's work. The activists described their new organizations as "grassroots" to distinguish themselves from the nationals. In the context of other social movements, the term is often associated with mass participation, but this was not the case for most grassroots biodiversity groups. They found that a small handful of determined activists could achieve significant environmental protection through litigation without having to engage in a broad mobilization of the public. (Furthermore, these groups had little chance of building a large membership base through direct mail because that terrain was already dominated by the national environmental organizations, as I will discuss later.) Instead, for the new groups, grassroots indicated that they were rooted in site-specific struggles over forest and wildlife protection and, unlike the nationals, they did not rely on an insider strategy focused on appealing to politicians in Washington, DC.

Because the grassroots biodiversity groups used lawful tactics, unlike Earth First!, they were eligible to receive tax-deductible contributions from grantmaking foundations. With this funding, the new groups could now provide modest salaries for their founders. And as they accrued litigation victories and attracted more funders, they were able to hire a few additional staff members. They often drew them from the Earth First! community, employing activists who were talented and capable, but who did not have the professional credentials, political connections, and insider mindset to be hired by the national organizations. For the first time, a sizable number of radical activists were able to work full time on biodiversity protection outside of the constraints of the moderate national organizations.

The growth of the grassroots biodiversity groups was aided by new funding sources, Earth First!'s movement culture, and the political conditions of the 1990s. Foundation grants were crucial to the growth of these groups. If they had instead tried to rely on fundraising through direct mail like the nationals, they would have encountered daunting obstacles. As noted earlier, direct mail requires extensive up-front funding. Moreover, the large national organizations already dominated direct mail fundraising on wildlife and wildlands protection issues. The small grassroots groups had little hope of competing with them on this terrain. Foundations offered an alternate source of funding that lowered the barriers-to-entry for the new grassroots groups.

The significance of foundation funding for the growth of the radical grassroots groups is ironic. Foundations have often been seen as a moderating force on social movement advocacy, and most large foundations do not fund radical groups.[46] However, the grassroots biodiversity groups were able to at least partially circumvent these constraints because of new environmental funders that appeared in the late 1980s. This was a time when environmental issues were receiving a great deal of public attention, stimulated by events such as the Chernobyl nuclear power accident, the *Exxon Valdez* oil spill, accelerated loss of rainforests, the discovery of a hole in the ozone layer, and culminating in the twentieth anniversary of Earth Day in 1990. Amid this widespread interest in environmentalism, philanthropic foundations were more likely to incorporate environmental issues into their grant programs.

The influx of new environmental grantmakers brought increased opportunities for funding alternative approaches within the environmental movement. As Peter Galvin of the Center for Biological Diversity explained,

> When you think about the new generation of funders that came online in the late '80s and early '90s, it really was a dramatic shift. If you look at conservation as investment, there is this sense of the big national groups as being these "blue chip" groups that everybody would put their investment in. But then as new people started to come into money, they said, "Let's try something different." And at that time, there was an explosion of smaller environmental groups forming.

In my interviews with grassroots biodiversity activists, a few of the newer environmental grantmakers were frequently mentioned as providing crucial early support for their groups. Two in particular were the Foundation for Deep Ecology (FDE), created by Esprit Clothing founder Doug Tompkins in 1990, and the Turner Foundation, begun by media mogul Ted Turner in 1991. Galvin recalled, "I would say that FDE and Turner at that time were probably the two most significant players in getting what some people have called the 'new conservation movement' off the ground."

These foundations hired program officers who had previously been involved with environmental organizations that did not rely on an insider strategy. Turner Foundation's environmental program was directed by Peter Bahouth, the former executive director of Greenpeace USA. The Foundation for Deep Ecology brought in Bill Devall and John Davis from Earth First!. Devall was a leading proponent of deep ecology and had administered the Earth First! Foundation/Fund for Wild Nature. John Davis had been the editor of the *Earth First! Journal*. Another notable new funder of grassroots groups was the outdoor equipment company Patagonia, whose environmental grant program was staffed in the second half of the 1990s and early 2000s by John Sterling; Sterling had previously worked for Earth Island Institute, an umbrella organization for grassroots environmental groups. Similar developments were evident within a few of the more established environmental grantmakers as well. For example, the Levinson Foundation had been involved in the formation of the Group of Ten, but in the 1990s its environmental program was directed by Charlotte Talberth, who had previously been active in an Earth First! offshoot called the Cathedral Forest Action Group (see chapter 4).[47] Because of their backgrounds, these new foundation program officers were more likely to be supportive of radical activists and as a result they played crucial roles in fostering the development of the grassroots biodiversity groups.

The grassroots biodiversity groups also benefited from the movement culture of Earth First!, especially in their early years. As noted earlier, many of the founders of biodiversity protection groups were shaped by Earth First!'s movement culture, which nurtured their use of an outsider strategy. The Earth First! community also provided the training ground for a pool of dedicated radical activists that these groups could draw from as they added more staff. Facets of Earth First!'s movement culture aided the grassroots biodiversity groups in other ways as well. For example, the *Earth First! Journal* was an important medium for these groups to share their experiences and learn from each other. Before the growth of Internet-based communications, there was no other comparable space for dialogue within the environmental movement. Likewise, the Round River Rendezvous provided an annual opportunity for grassroots activists to reconnect and reaffirm their outsider approach. Not all grassroots activists participated in Earth First!, but because most grassroots biodiversity groups did not develop significant community-building institutions of their own, Earth First!'s movement culture provided an important resource to help sustain the radical approach of the new groups in the 1990s.

The political conditions during the 1990s also aided development of the grassroots biodiversity groups. While grassroots groups were able to make some advancements under President George H. W. Bush in the early 1990s, the Clinton presidency proved to be particularly useful for helping the grassroots groups to distinguish themselves from the national environmental organizations in two ways. First, a Democratic administration was less likely to publicly support efforts to roll back environmental protections and would offer at least some pro-environmental measures. Environmentalists could thus spend less time on the defensive and instead be more proactive. Under these circumstances, grassroots groups were able to distinguish themselves from the better-known national organizations because the initiatives put forward by the nationals were more limited, based on a more cautious interpretation of political reality and concern over asking too much from their allies in the Democratic Party. By comparison, grassroots groups with no ties to politicians were free to advance comprehensive proposals that were based more on ecological science than political calculus.

Second, President Clinton was not a strong champion of environmentalism, particularly during his first term. However, because of their insider strategy and their ties to the Democratic Party, the national environmental organizations were reluctant to strongly criticize the Clinton administration when its environmental policies were inadequate or even detrimental. Instead, the grassroots groups often led the charge. As such, they appeared to be more consistent advocates for the needs of forests and wildlife than the nationals. Thus, both the opportunities and shortcomings of the Clinton administration made it easier for the grassroots biodiversity groups to distinguish themselves favorably from the national organizations and thereby garner broader support.

In 2001, the inauguration of George W. Bush brought in a Republican administration that was actively hostile to the environmental movement. Under these conditions, both the nationals and the grassroots groups had to spend more time fending off a rollback of environmental protections rather than promoting new initiatives. Likewise, insider environmental organizations linked to the Democratic Party became eager critics of a Republican administration. Playing defense against a Republican administration, the distinctions between the nationals and the grassroots groups became more muted.

While the political conditions became more challenging in the 2000s, by that time the grassroots biodiversity groups had already grown to become an important alternative to the national organizations in the realm of environmental advocacy, as illustrated by the case studies in the next three chapters.

3

NEVER MIND THE NATIONALS:
THE HEADWATERS FOREST CAMPAIGN

The Headwaters Forest campaign worked to save the largest remaining stands of old-growth redwood trees on private lands in northern California. Most of the dominant national environmental organizations played little or no role in this campaign. Instead, a network of grassroots groups with few resources was able to make Headwaters Forest into a national issue and compel the federal government to intervene to protect the redwoods. The earliest champion of Headwaters was Earth First!. During the Headwaters campaign, Earth First! underwent substantial tactical changes from focusing on clandestine monkeywrenching in the 1980s to a mass-based civil disobedience approach in the 1990s. The Headwaters campaign provided a key impetus for these developments and also offered one of the clearest expressions of the changes in Earth First! in the 1990s. In addition to Earth First!, the Headwaters campaign included an array of small environmental organizations. The most influential was the Environmental Protection Information Center. Its effective use of litigation, despite having scant resources, would serve as an early role model for subsequent grassroots biodiversity groups. The Headwaters campaign also fostered the emergence of many new groups that used a variety of other innovative tactics. The Headwaters Forest campaign was thus both a starting point and source of inspiration for the proliferation of grassroots biodiversity activism around the United States in the late 1980s and 1990s. And it offered a unique opportunity to see what grassroots groups could accomplish on their own without the national environmental organizations.

Redwoods, Earth First!, and the Origins of the Headwaters Forest Campaign

Redwood trees are renowned for their spectacular size. The coast redwoods are the tallest trees in the world and they can live for more than 2,000 years. Some old-growth redwoods reach heights exceeding 350 feet, with a diameter of 20 feet or more. The only species of tree that is larger in total mass is a close relative known as the giant sequoia, which is found in the Sierra Nevada mountain range of eastern California. Coast redwoods, by comparison, grow near the Pacific Ocean in northwestern California up to the Oregon border. This forest ecosystem depends on coastal fog to provide moisture and moderate temperatures, so the redwood belt extends only about 40 miles inland. Historically, this redwood zone covered 2 million acres. However, by the late twentieth century, more than 95 percent of the original forest had been logged.[1] Most of the remaining old-growth redwood groves were protected in California state parks and Redwood National Park. These groves were generally rather small—from a couple of hundred to a couple of thousand acres in size—offering roadside viewing opportunities for tourists, but they were not large enough to provide fully functioning ecosystems.

Despite more than a century of intensive logging, in the 1980s, some old-growth redwoods could still be found on private lands, most notably on the property of the Pacific Lumber Company in Humboldt Country. In 1986, Pacific Lumber (PL) owned about half of the remaining old-growth redwoods on private land.[2] For more than a century, the company had operated largely free of controversy and had even earned praise from some environmentalists for the comparatively low impact of its logging practices. PL relied primarily on selective logging rather than clearcutting, and its logging rates were low enough that a significant number of old-growth trees remained on its land. Although its workers were not unionized, the company was also known for its extensive employee benefits, including a large pension fund. However, the same attributes that made Pacific Lumber a respected member of its community also made it a prime target for an outside takeover. And, indeed, in 1986 the Maxxam Corporation took over Pacific Lumber. Maxxam's CEO, Charles Hurwitz, had a reputation as a corporate raider who would buy out other companies through money raised with "junk bonds" and then quickly sell off their most valuable assets. The high interest rates on the $754 million in junk bonds used for Maxxam's takeover of Pacific Lumber meant that Hurwitz would need to liquidate PL's assets as quickly as possible in order to repay this debt.

The Maxxam takeover caught the attention of Greg King, a local journalist and environmental activist. King reviewed Maxxam's shareholder prospectus, which revealed the corporation's intention to double PL's logging levels and shift to clearcutting. King contacted Darryl Cherney, a musician who had organized the first Earth First! action in the region a few months earlier. Together they called a meeting in which the Earth First! Redwood Action Team was formed.

By October 22, 1986, Earth First! had organized the first protest of what would become the Headwaters campaign, outside of Pacific Lumber's office in San Francisco. The Redwood Action Team began to monitor Maxxam's timber harvest plans and would trespass on Maxxam's property in order to examine areas scheduled to be logged. It was during one of these explorations in late March 1987 that Greg King discovered the largest stand of old-growth redwoods remaining on private lands. The main grove was over 3,000 acres, making it comparable to some of the largest protected groves in state parks. But unlike the state parks, this grove had not been significantly impacted by human visitors, so it retained a primeval character, with 10-foot-tall ferns and abundant native wildlife. Because of its location at the beginning of two rivers, King named it "Headwaters Forest." Headwaters Forest would become the centerpiece of the campaign against Maxxam.

Soon thereafter, in May 1987, Earth First! organized a National Day of Outrage against Maxxam, which involved demonstrations in five locations in four states. It was the first time that Earth First!ers were arrested for protesting Maxxam. This was followed by the introduction of the tactic of tree-sitting to northern California. Tree-sitting entailed an Earth First!er perching on a small platform high up in a tree that was scheduled to be cut; the tree could not be cut without injuring or killing the tree-sitter. Tree-sits would generally last no longer than a few days or weeks, during which time the activists sought to forestall the logging and bring attention to the issue. The tree-sits and other forms of direct action attracted extensive media coverage of Maxxam's impact on the redwoods, including a *Reader's Digest* article titled "California's Chain-Saw Massacre."[3]

Beyond Earth First!, concern over Maxxam's takeover extended to Pacific Lumber workers. In addition to liquidating the forests, Hurwitz raided their pension plan to help pay off his debts. Five hundred PL employees signed on to a full-page ad objecting to Hurwitz's takeover. Hurwitz's reply to employees' criticism was to cite the golden rule: "He who has the gold rules."[4] The workers' frustration with Hurwitz's actions created the possibility of Earth First!ers and timber workers finding common cause against Maxxam.

The activist most identified with building this alliance was Judi Bari. Bari was a carpenter with a background of union organizing and radical political activism, that shaped her proworker approach to environmental activism. In addition to being an Earth First! activist, she cofounded a local chapter of the International Workers of the World to support timber workers in situations where the workers' unions were unable or unwilling to do so. Through this local chapter, Bari was able to advocate on behalf of timber workers who had been exposed to a toxic chemical spill at a timber mill when the workers' union refused to help them. Bari also campaigned on behalf of the widow of another mill worker who had been crushed by a load of timber after being pressured by his supervisor to clear jams in the assembly line without stopping the machinery. Bari won partial victories in both of these cases and established a reputation as an ally to timber workers. Bari's alliance-building stood in stark contrast to the image of environmentalists versus workers that was being promoted by the timber industry.

An important breakthrough in Bari's outreach to timber workers revolved around the issue of tree-spiking. Tree-spiking was the most famous type of monkeywrenching done by Earth First! in the 1980s. Indeed, it was a central part of Earth First!'s image. Tree-spiking entailed driving long nails into trees that were scheduled to be logged. If a spiked tree was cut and taken to the mill, the spike would likely hit a saw blade, causing extensive damage to the mill equipment and potentially injuring mill workers as well. Causing harm to timber workers was not a goal of this tactic; Earth First!ers sent out announcements to warn of where they had spiked trees. The goal was to provide an economic deterrent where the risk of damage would cause timber companies to forgo trees in areas that had been spiked. However, there remained the risk that timber companies would still cut trees in a spiked area and not successfully remove all the spikes, so timber workers felt imperiled by tree-spiking.[5] It was a source of resentment from workers toward Earth First! and a key barrier to any alliance between them. As a result, in March 1990, at the urging of a millworker named Gene

Lawhorn, Bari publicly renounced tree-spiking. Soon after, Earth First! activists in northern California and southern Oregon declared that they would cease tree-spiking in their regions.

While Bari and other Earth First!ers were making progress in their outreach to workers, the attention garnered by their organizing was in turn accompanied by a growing number of incidents of violence against them. Judi Bari, Darryl Cherney, and Greg King had been assaulted and had received death threats, yet they found that the local police and media largely turned a blind eye to this violence. In response to the climate of intimidation, the activists decided to apply the tactics from the civil rights movement's 1964 Mississippi Freedom Summer campaign to an environmental context with an event they called "Redwood Summer." Their goal was to bring thousands of college students and other supporters to northern California in the summer of 1990 to engage in nonviolent civil disobedience and garner national attention for the plight of the redwoods.

On May 24, 1990, amid the organizing for Redwood Summer, Bari and Cherney were driving through the city of Oakland when a pipe bomb exploded under Bari's car seat. The pipe bomb was wrapped with nails to produce shrapnel and Bari was critically injured by the blast. Her pelvis was shattered and she was crippled for life. Cherney escaped serious injury, but was arrested by Oakland police and FBI agents who quickly appeared on the scene. Bari was also placed under arrest while she was still in surgery. The Oakland police, in conjunction with FBI agents, wrote an affidavit for a search warrant declaring that "Bari and Cherney are members of violent terrorist group involved in the manufacture and placing of explosive devices," when in fact Earth First! had never used explosives.[6] The FBI and Oakland police's comments to the media fostered the impression that Bari and Cherney had been knowingly transporting explosives when the bomb accidentally detonated. For example, a *San Jose Mercury-News* article began, "Two members of the radical environmental group Earth First! were injured Thursday by their own pipe bomb."[7] However, it soon became apparent that the explosion had not been accidental and that the pipe bomb had been triggered by a motion-sensitive device. Bari and Cherney ultimately were never charged in the case, but the FBI did not retract its representation of them as the bomb makers. They would later sue the FBI and Oakland police for false arrest and civil rights violations, drawing parallels between their treatment and the FBI's programs against progressive activists in the 1960s and 1970s. At the time of the bombing, the negative publicity from the accusations dealt a serious blow to their emerging alliance with timber workers.

The bombing also initially jeopardized Redwood Summer, but while Judi Bari remained hospitalized into the summer, many women stepped forward to do the organizing for that event. Bari would later write, "Redwood Summer was the feminization of Earth First!, with 3/4 of the leadership made up of women."[8] Redwood Summer ultimately drew around 3,000 participants. (Prior to that time, the largest Earth First! actions in that region had involved only about 150 people.[9]) Participants were required to follow a nonviolence code that also forbade monkeywrenching. Instead, they engaged in mass civil disobedience, including numerous tree-sits and blockades of logging trucks. There were almost daily actions from June through August, with more than 250 arrests.[10] The largest event of the summer was a rally at a logging mill that drew more than 2,000 people, an unprecedented scale for an Earth First! action.

Redwood Summer marked a key shift in Earth First! away from a focus on monkeywrenching and toward the mass civil disobedience that would characterize Earth First!'s tactics in the 1990s. Accordingly, Earth First!ers in northern California did not have to operate under the level of secrecy that monkeywrenching had required. Instead, they were able to work more openly in the community. Although Earth First!ers still got arrested for engaging in civil disobedience, they were no longer associated with property damage, so other organizations became more comfortable working in coalition with Earth First! on the Headwaters campaign.

EPIC, the Sierra Club, and Litigation

Along with Earth First!, the other group most identified with the Headwaters campaign was the Environmental Protection Information Center. This organization—which its members usually referred to as "EPIC" rather than its full name—can best be understood in the context of its location and history. Not an urban-based environmental group, EPIC was formed in the small town of Garberville in Humboldt County. In a region long dominated by the presence of the timber industry, Garberville stands out as a counterculture sanctuary, reflecting the influx of the back-to-the-land movement of the 1970s. EPIC began in 1977 when a group of back-to-the-land activists organized to oppose aerial spraying of herbicides connected with logging near their homes. The group then got involved in the protection of a grove of old-growth trees adjacent to the nearby Sinkyone Wilderness.

As part of this campaign, EPIC began exploring litigation in 1983. In California, logging on private lands is regulated through timber harvest plans, which must be approved by the California Department of Forestry and Fire Protection (known simply as CDF). EPIC's first lawsuit against CDF over its approval of a timber harvest plan near the Sinkyone was successful in protecting that area, and it also had broader implications. The courts ruled that CDF had to consider the cumulative impacts of each timber harvest plan. Up until that time, the ecological harm from logging on private lands had been "death by a thousand cuts." CDF could contend that a single logging project in and of itself would not destroy a particular watershed, and thus would approve each plan. However, as a result of the lawsuit, the agency was required to look at the impact of a timber harvest plan not in isolation, but rather in terms of how it would contribute to the larger set of cumulative effects on the ecosystem from all of the logging in that area. EPIC's case created an important legal precedent for challenging future logging projects, and EPIC became the primary organization using litigation to challenge private lands logging in California.

The organization was able to accomplish these activities with very little funding. As a small group with few resources that was still able to have a big impact on environmental policy, it served as an important precursor for the wave of new grassroots biodiversity groups that emerged in the late 1980s and 1990s. EPIC did not get its first foundation grant until 1986 (which included a line item for the group to buy its first electric typewriter). Instead, for many years the group persisted through the local support of the Humboldt counterculture community, which nurtured grassroots environmental activism in the region. Another critical component was the role played by lawyers such as Sharon Duggan and Tom Lippe who volunteered to file much of the organization's early litigation for little or no remuneration.

Above all, EPIC was sustained for many years by the hard work of its all-volunteer staff. (No staff members received a salary until the 1990s.) Among the most visible founding members were Robert Sutherland, Richard Gienger, and Cecilia Lanman. Sutherland (who went by the name "The Man Who Walks in the Woods") taught himself law and then prepared the legal brief for EPIC's groundbreaking cumulative effects case. The group's activities also extended beyond litigation. Gienger led a program

of intensive monitoring of timber harvest plans and the state agencies reg-
ulating those plans. Lanman was representative of EPIC's public education
work. She had come to Humboldt County to homestead and prior to that
had been involved with the United Farm Workers. She would become the
EPIC organizer most identified with the Headwaters campaign and was
EPIC's executive director through the height of the campaign.

EPIC got involved in the Headwaters campaign just as its work to protect the Sinkyones was being completed. At that point, Darryl Cherney
was on EPIC's board of directors and encouraged the group to take up the
Maxxam issue. Soon thereafter, Cherney stepped down from the board when
the local Earth First! activists began engaging in civil disobedience against
Maxxam. In order to conduct its litigation, the group needed to maintain
"clean hands" in the eyes of the courts. In other words, EPIC might lose
its standing to bring a lawsuit against a timber harvest plan if the group was
involved in illegal activity protesting that logging. Therefore it had to
maintain formal separation from Earth First!. However, the two groups continued to work together closely wherever possible. Indeed, some of EPIC's
key organizers during the Headwaters campaign first got involved with the
issue through Earth First!. And, unlike most national environmental organizations, EPIC did not criticize Earth First! in the media.

EPIC's first lawsuit against Maxxam was filed in 1987. This suit involved
clearcutting planned for the heart of Headwaters Forest. In this case, the
judge found that the California Department of Forestry had "rubber
stamped" approval of the timber harvest plan without sufficient analysis
and had prevented the staffs of other state agencies from making comments critical of the plan. Over the next decade, EPIC filed thirteen lawsuits specifically over Maxxam and Headwaters and was successful in more
than 60 percent of these cases.[11]

EPIC also helped to bring the Sierra Club into the Headwaters campaign. In 1988, the Sierra Club joined in EPIC's third lawsuit against
Maxxam. The Club's role in the Headwaters campaign was complicated,
reflecting its dual character as a staff-based national environmental
organization on the one hand, but simultaneously having an active local
membership component organized through a system of regional chapters, with the Redwood Chapter covering coastal northern California.
For much of the Headwaters campaign, it was the chapter-level volunteers, rather than the national-level staff, who were the primary Club
activists on this issue.

The local Sierra Club activist who played the most prominent role in the Headwaters campaign was Kathy Bailey. Like Cecilia Lanman, Bailey came to this issue with a background in the protest movements of the late 1960s and early 1970s. She then moved to northern California and got involved in a campaign against aerial herbicide spraying of forests. This experience led to her interest in forest issues and participation in the Sierra Club. She became the Club's primary representative in the Headwaters campaign from 1993 through the rest of the decade. However, she did so as a full-time volunteer and personally had to do much of the fundraising to support the Club's participation in the campaign.

Forests Forever and the Grand Accord

While EPIC and the Sierra Club were having success in the courts, the process of challenging the logging on a project-by-project basis was extremely demanding, especially for a small volunteer organization like EPIC. Its activists began looking for an alternate strategy to piecemeal timber harvest plan litigation. In 1989, they organized a meeting that brought together grassroots forest protection activists to explore using California's ballot initiative process as means to directly increase protection for private forestlands. If the supporters of an initiative gathered 600,000 signatures from registered voters, the measure would be added to the ballot in the next statewide election. If it then received a majority of votes, it would become law. Therefore, this approach could sidestep the shortcomings of the state regulatory agencies and the grassroots activists' limited influence with state politicians. To pursue this approach, the grassroots activists formed a new organization called Forests Forever and began crafting the language of an initiative that would share its name.

The Forests Forever initiative had two primary components. One part was a $710 million dollar bond issue for the acquisition of remaining old-growth groves on private lands. Headwaters Forest was specifically named in the text of the initiative as a priority for acquisition. The second part was an extensive reform of the state forestry regulations. These reforms would sharply curtail clearcutting, increase protection for wildlife impacted by logging, protect trees along fish-bearing streams and rivers, and develop a new sustained yield system for commercial logging throughout the state.

The initiative attracted the interest of a wealthy donor named Hal Arbit, who provided millions of dollars to bankroll it. The initiative also drew in larger state and national environmental organizations, including the staff from the Sierra Club, Natural Resources Defense Council, and Planning and Conservation League (the California affiliate of the National Wildlife Federation). But the impact of the influx of money and the involvement of the national organizations would create friction with the grassroots activists in a pattern that would appear again later in the Headwaters campaign. The Sierra Club staff insisted that the initiative be rewritten as condition of their participation, and Arbit used his financial influence to exert more control over the campaign process as a whole. The result was a dual structure within Forests Forever. As Cecelia Lanman summarized, "There was the professional steering committee that had the bigwigs on it that allowed us to bring in enough funding to make it a viable campaign. And then there was the original [grassroots] Forests Forever which held the line on compromise."

The Forests Forever initiative was successful in getting enough signatures to qualify for the November 1990 ballot and was designated Proposition 130. At that time, it enjoyed strong support in public opinion polls, but it ultimately wound up losing by a narrow margin, garnering 48.5 percent of the vote. For grassroots activists, the control that the professional environmentalists supported by Arbit exerted over the Forests Forever campaign was highlighted as a central factor in the initiative's defeat. The primary strategy of Forests Forever's opponents was to try to tar that initiative through association with Earth First!. There was a sharp division between the professional and grassroots wings of the Forest Forever campaign in their responses to the opposition's strategy. The response from the professional wing was to further distance themselves from Earth First!. In so doing, they alienated important grassroots organizers within the campaign. Kathy Bailey contended, "The reason we lost Prop. 130 is because the original campaign manager was so worried about our image and that our image would be inextricably linked with Earth First!."

Grassroots activists were also critical of the role that Arbit played in steering the Forests Forever campaign. Writing in 1991, Gary Ball of the Mendocino Environmental Center asserted that the initiative lost "largely due to the inept campaign put on by Arbit's hand-picked campaign team. Hal Arbit, you see, has a hard time taking advice from grass roots folks or ordinary people."[12] Likewise, Darryl Cherney of Earth First! highlighted Arbit's purse strings as the reason Forests Forever failed. He was particularly critical of how Arbit's money was allocated. According to Cherney's estimate, out of $7 million spent on the Forest Forever initiative, only about $100,000 was spent on grassroots organizing, while much of the rest was spent on expensive television and radio ads. Cherney illustrated the contradictory role played by Arbit's money when he declared, "It is the reason [Forests Forever] almost passed, and it is the reason it failed."

Despite these obstacles, the Forests Forever initiative came quite close to victory. It garnered more votes than other environmental ballot initiatives in California that year that had been created by large, moderate environmental organizations. Many grassroots activists looked forward to reentering the initiative in next election. However, Arbit instead focused on negotiating a deal with the timber industry. By early 1992, the negotiations between the large environmental organizations supported by Arbit and some timber companies produced the "Grand Accord," a compromise forestry measure that would need to be finalized through state legislation. Over the objections of the grassroots, Arbit refused to file the signatures that had been gathered to put the Forests Forever initiative back on the ballot, and Arbit instead promoted the Grand Accord legislation.[13] Grassroots activists generally condemned the Grand Accord, which they saw as conceding too much to the timber industry and providing far less protection than the initiative. In a newspaper editorial on the subject, Kathy Bailey wrote, "Many in Sacramento believe we should accept a bad deal rather than get no deal at all—take the couple of inches of grimy bathwater and forget there was ever a baby."[14] Bailey worked to get the Sierra Club to withdraw its support for the accord and to stop it. However, the accord was still endorsed by the Planning and Conservation League, Natural Resources Defense Council, and Audubon. One key politician who opposed the accord was Tom Hayden, a prominent student activist in the 1960s who had later been elected to the California State Assembly. Writing to the speaker of the assembly to express his opposition to the accord, Assemblyman Hayden ended his letter with a quote from John Muir: "This playing at politics saps the foundations of righteousness." Ultimately, in the face of these divisions, the Grand Accord was defeated.

Focus on Headwaters Forest

At the same time that the Grand Accord was unraveling, the year 1992 also saw four developments that refocused the attention of environmentalists on the ancient redwood groves of Headwaters Forest. The first development was the U.S. Fish and Wildlife Service's designation of the marbled murrelet as a threatened species under the Endangered Species Act in October 1992. The marbled murrelet is small seabird that nests on the large branches of coastal old-growth trees. Thus the fate of the murrelet was tied to the giant redwoods of Headwaters, much as the more famous northern spotted owl depended on the old-growth forests of the Pacific Northwest (see chapter 4). The protection of the murrelet under the Endangered Species Act shifted the regulatory oversight of logging around Headwaters Forest. Whereas before it had been the sole domain of California state agencies, now a federal law enforced by a federal agency could play a role in the fate of Headwaters. This shift also created new litigation opportunities for EPIC at the federal level. EPIC would quickly find the basis to apply them, following the second major development of 1992, known as the "Owl Creek Massacre."

The month after the murrelet was listed, over Thanksgiving weekend, Maxxam sent a team of loggers to cut Owl Creek Grove, one of the old-growth groves comprising Headwaters Forest. Since 1989, EPIC and the Sierra Club had blocked Maxxam's efforts to log in Owl Creek, and Maxxam's Thanksgiving logging was illegal.[15] However, with the holiday weekend, EPIC could not get a court injunction to stop the logging immediately, and no regulatory agency or law enforcement agency would intervene at that point either. It was left to the Earth First! activists to place their bodies between the chainsaws and the trees. One account of the action began with the following exchange:

> "You better stay away from my chainsaw!" shouted an angry old-growth faller.
>
> "My hand is on the bar of your chainsaw, and you better not start it 'cause my hand will be ripped up," Earth First!er John Garcia rebutted. "Take this day off, please, and I'll invite you to dinner."[16]

Despite such daring actions, Maxxam was able to cut and remove a million dollars worth of old-growth redwoods from within the grove. Based on this damage, EPIC turned to the federal courts and the Endangered Species Act to challenge Maxxam's harm to marbled murrelet habitat from the Owl Creek logging. Attorneys Mark Harris and Macon Cowles stepped forward to take this large case for EPIC pro bono. They prevailed in 1995. The ruling set a precedent for protecting the old-growth groves. This case (*Marbled Murrelet and EPIC v. Pacific Lumber*) was the first time that the

Endangered Species Act had been applied to stop logging on private land in order to protect an endangered species' habitat.

The third notable development of 1992 was the Headwaters slideshow created by Doug Thron, a native Texan who had moved to northern California to attend college. There he began regularly trespassing on Maxxam's property in order photograph Headwaters and then turned these images into a popular slideshow. The slideshow proved to be an important organizing tool that planted images of Headwaters in the public's consciousness. Thron's photographs made a place that was secluded behind private property lines tangible to audiences who could not otherwise see Headwaters directly. A year after the Owl Creek Massacre, Thron took his slideshow on the road to audiences across the United States. As with much of the grassroots organizing in the Headwaters campaign, it was accomplished with few resources, with Thron relying on donations at each location to provide the gas money to get him to the next event. The success of this outreach led *Time* magazine to juxtapose Thron with fellow Texan Charles Hurwitz in a feature on Headwaters Forest.[17] The article was illustrative of the way that the campaign against Maxxam had become increasingly focused on Headwaters.

The fourth key development of 1992 was the election of Congressman Dan Hamburg (D-CA), who would introduce the Headwaters Forest Act. In November 1992, Hamburg was elected to the congressional seat representing coastal northern California. Up until this time, the elected officials representing the region usually had ties to the timber industry, regardless of whether they were Republicans or Democrats. By contrast, Hamburg was an outspoken critic of the timber industry. The 1992 elections appeared to create a promising political opportunity for the Headwaters campaign. Not only had Hamburg been elected, but there was also now a Democrat as president for the first time in twelve years. President Clinton had actively courted environmentalists during his campaign. Moreover, from 1993 through 1994, the Democrats also had sizable majorities in both houses of Congress. The opportunities for a federal legislative solution to Headwaters seemed strong.

After taking office, Hamburg was besieged by constituents calling his attention to the plight of Headwaters, and he soon took on the protection of Headwaters as a primary legislative goal. Local forest activists were pleased to find that he did not seek to distance himself from grassroots groups or even from Earth First!ers. Indeed, they were actively involved

in crafting Hamburg's legislation. For example, Judi Bari got the Headwaters legislation to include funding for ecological restoration jobs for displaced timber workers. And while members of Hamburg's staff initially
sought to limit the scope of the bill solely to the 3,000-acre Headwaters
Grove, the activists were successfully able to push for language calling for
the acquisition of 44,000 acres encompassing the greater Headwaters Forest ecosystem.

The Headwaters Forest Act was introduced in the House of Representatives in August 1993. The bill was approved by the House in September 1994 by a vote of 288 to 133. However, Hamburg and the Headwaters
activists had a difficult time getting the Senate companion bill introduced. In order to be taken seriously, the bill needed to be sponsored by
one of California's U.S. senators. Both senators—Dianne Feinstein and
Barbara Boxer—were Democrats. Boxer was generally seen as the more
liberal of the two and the logical champion for a Senate bill. However,
she seemed unwilling to take it up. It was not until after Earth First! organized a protest outside of Boxer's San Francisco office that she introduced
the Senate companion bill in July 1994. However, that step apparently
marked the end of her efforts on behalf of the Headwaters Forest Act, and
Darryl Cherney placed the blame for the subsequent failure of the legislation on Boxer. Indeed, while Hamburg had rallied 142 representatives
to cosponsor his bill, Boxer's Senate version ended the year with only one
cosponsor. The Headwaters Forest Act died when the 103rd Congress came
to a close without having the Senate version of the bill come to a vote.

By that time, it was also clear that Rep. Hamburg would not be back
to reintroduce his legislation. He had been defeated by Republican Frank
Riggs in November 1994 after having also faced a well-funded challenge
in the Democratic primary from Doug Bosco, a former congressman who
had subsequently been hired as a lobbyist by Hurwitz. Hamburg left office
deeply disappointed with the Clinton administration and the Democratic
Party. He became a strong critic of both and joined the Green Party. His
experiences in Washington also led him to question the value of the sort
of "access" coveted by insider environmental groups. He later recounted
his own level of access as a congressman:

> I was able to sit down with Clinton on Air Force One and make the
> case for Headwaters. But then he'd go golfing with Vernon Jordan [a
> Clinton adviser who was hired by Hurwitz]. You end up getting hood-

winked. You think you're really making a lot of progress because
you're standing around eating shrimp and drinking Bloody Marys with
all the higher ups, but in reality they're screwing you over.

After leaving Congress, Hamburg remained active in the Headwaters campaign through his work in the advocacy group Voice of the Environment.

New Organizations and Tactics

While the Hamburg bill ultimately did not prevail in Congress, the existence of this legislation fostered a surge in Headwaters activism and the
involvement of new organizations that brought a diversity of innovative
tactics to the campaign. In particular, the prospect of a legislative solution to the Headwaters issue offered a more prominent role for organizing
work in urban areas, especially in the San Francisco Bay Area. By the end
of 1993, Headwaters activists there formed the Bay Area Coalition for
Headwaters (BACH) to coordinate their efforts. Despite the name, rather
than being a formal coalition, BACH provided open space for coordination among a diverse group of urban activists concerned with Headwaters. The ad hoc assemblage ranged from Earth First!ers to Sierra Club
members and participants from various walks of life. Many of the principal organizers in BACH, as within the rest of the Headwaters campaign,
were people in their forties who had been notably influenced by the protests
of the 1960s during their youth and had participated in other social movements in the 1970s and 1980s.

The activist most identified with BACH was Karen Pickett. Pickett
had been involved in the Maxxam/Headwaters campaign from its inception through her role in the Bay Area Earth First! group. In Humboldt,
Earth First! organizers could subsist on very little money by living in back-
woods homesteads off the grid. Pickett managed to carve out a similar space
for herself amid the expensive Bay Area real estate. She lived in a cabin
in a tiny rural enclave amid the urban sprawl, with an outhouse and water
supplied by leaky tubes connected to a nearby spring. Living simply was
requisite for the BACH organizer, especially in the early days when BACH
had no foundation grants, no paid staff, and no office.

Much of BACH's crucial initial financial support came from another
of its cofounders, Ken Miller. Miller worked as an emergency room doctor, a job that provided the flexibility for him to pursue his Headwaters

activism and was lucrative enough that he could serve as BACH's bene-factor in its early years. Then in 1997, he moved to Humboldt County to help start a variety of organizations including Taxpayers for Headwaters, Salmon Forever, and the Humboldt Watershed Council. These groups worked to organize local communities harmed by the effects of Maxxam's logging in the region.

Michael Passoff was BACH's first coordinator. Passoff embodied the diversity of organizations and tactics involved in the Headwaters campaign. In addition to his role in BACH, he was also chair of the ancient forest committee of the San Francisco Bay Chapter of the Sierra Club, whose volunteers were active in the Headwaters issue. Then in 1996, Passoff was hired to coordinate a boycott of old-growth redwood for Rainforest Action Network. Since Maxxam was the main purveyor of old-growth redwood, this boycott put direct financial pressure on the corporation, especially after the campaign succeeded in getting large retailers of redwood such as Home Depot to eschew old growth. The redwood boycott set an important early precedent for market-based tactics within the forest protection movement. After that, Passoff worked at the As You Sow Foundation, a group that organized shareholder activism on environmental and social justice issues. The As You Sow Foundation sponsored shareholder resolutions urging Maxxam to be a willing seller of Headwaters Forest, as well as running reform candidates for Maxxam's board of directors.

Another organization involved in the Headwaters shareholder activism was the Rose Foundation for Communities and the Environment, but the Rose Foundation became best known for its work using another tactic.[18] Soon after starting the Rose Foundation in 1993, attorney Jill Ratner read about a tactic that Darryl Cherney of Earth First! was promoting to protect Headwaters called "debt-for-nature." The premise for debt-for-nature was Charles Hurwitz's role in the bankruptcy of a savings and loan he controlled—the United Saving Association of Texas. It was the fifth most expensive savings and loan failure in U.S. history, and the federal bailout cost taxpayers $1.6 billion.[19] The Federal Deposit Insurance Corporation contended that financier Michael Milken had underwritten Hurwitz's takeover of Pacific Lumber in return for the United Savings Association of Texas investing in Milken's high-risk junk bonds, resulting in the savings and loan's bankruptcy. In 1988, the Federal Deposit Insurance Corporation had charged one of Hurwitz's organizations $548 million for its role in the savings and loan's failure, but Hurwitz challenged this fine. At the time, another federal agency, the Office of Thrift Supervision, was also investigating the issue. It was unclear whether Hurwitz would be able to pay for the potential fines from both regulatory agencies without bankrupting Maxxam. Cherney promoted the idea that Hurwitz could instead give the federal government Headwaters Forest lands equivalent in value to whatever settlement might result from the savings and loan case.

With her background in corporate law, Jill Ratner was intrigued by the debt-for-nature approach, and with Cherney's blessing, the Rose Foundation took the lead on this tactic. This development illustrated how the Headwaters campaign offered a space for innovative collaborations between Earth First! and more conventional groups. Debt-for-nature was also indicative of the complementary roles played by the growing array of tactics in the Headwaters campaign. Indeed, soon after Boxer introduced the Senate companion to the Hamburg bill in July 1994, a journalist used it to illustrate "how the parallel efforts by preservationists are converging. The [EPIC] injunction sought on behalf of the marbled murrelet, if successful, will lock out the loggers while Congress deliberates [on the Headwaters Forest Act]. And as the preferred method of funding her legislation, Boxer cites a debt-for-nature swap."[20]

The proliferation of new Headwaters organizations in the San Francisco Bay Area was matched by a similar growth in Humboldt County. At the heart of this proliferation of Humboldt groups was the Trees Foundation. It began in 1991 as a vehicle for distributing grants from a local wealthy donor to environmental activists in Humboldt County, but it grew into a broader role as a "nonprofit incubator" for forest protection organizations in the region. It served as a fiscal sponsor for emerging new groups and provided a variety of services to foster their growth, including administrative tasks, design and publications, organizational development, trainings, and technical support. Many Headwaters-related organizations were the beneficiaries of the Trees Foundation's support, such as the Headwaters Action Video Collective; Institute for Sustainable Forestry; Mattole Restoration Council; Native American Coalition for Headwaters; Darryl Cherney's music performances and event organizing through Environmentally Sound Promotions; and public educational activities by Earth First!; as well as other notable projects that will be discussed later, such as the pepper spray lawsuit and Julia Butterfly Hill's Circle of Life Foundation. The Trees Foundation also served as the hub for the coalition of Headwaters groups known as the Headwaters Forest Coordinating Committee, which formed amid the growing level of activity in 1995.

The September 15th Rally of 1995

In 1995, Maxxam renewed its efforts to log Headwaters Grove. The combination of EPIC–Sierra Club litigation and the prospect of government purchase of Headwaters had kept Maxxam at bay for years, but with the demise of the Hamburg legislation, the likelihood of a buyout seemed dim. At the same time, environmentalists were making headway with the favorable federal court ruling in the Owl Creek case in February 1995 to protect murrelet habitat within Headwaters Forest from logging. Maxxam responded with a twofold challenge to the protection of Headwaters.

One response was a "takings" claim against the federal government in which Maxxam argued that if the Endangered Species Act prevented Maxxam from cutting down old-growth trees that were critical habitat for the marbled murrelet, then the federal government was taking Maxxam's property without compensation in violation of the Fifth Amendment of the Constitution. Maxxam contended that the federal government should either eliminate the protection of the murrelet habitat or else pay Maxxam the full value of the trees it could not cut. During this time, opponents of endangered species protection frequently raised the takings issue, though it primarily served a rhetorical role and had not resulted in any significant courtroom victories.

Second, just three days after the Owl Creek ruling, Maxxam took advantage of a loophole in the California state forestry rules. A "salvage" exemption allowed the company to cut 10 percent of its forests without any environmental review under the pretext of removing dead, dying, or diseased trees—a loose description that could be applied to almost any tree. Faced with the prospect of salvage logging in Headwaters, in March 1995 Earth First! held a rally that attracted about 500 community members and organized direct action to hinder logging. In response, Maxxam agreed to delay their plans to log in Headwaters. At that point, Maxxam was restricted by the Endangered Species Act from logging in murrelet-occupied areas until the nesting season officially ended on September 15th.

The nesting season gave Headwaters activists a respite in which to mobilize to stop this new threat. EPIC prepared litigation to challenge the legality of Maxxam's salvage logging. Meanwhile, Earth First! organized for a rally and direct action on September 15th, the day before Maxxam's logging was scheduled to resume. The event would take place in the tiny mill town of Carlotta at the gate of Maxxam's logging road leading to Headwaters Forest. The Earth First!ers decided to organize this action in a manner that would encourage much broader participation. Mobilizing Headwaters supporters from the urban centers around San Francisco to make the five-hour drive north would be one important component. The Earth First!ers also made a conscious effort to foster an environment where non–Earth First! activists could feel safe participating, not only in the rally, but also in the civil disobedience. This goal required them to create a structured situation in which large numbers of people could engage in civil disobedience without being injured by police and without the unplanned arrest of the rally participants who were not taking part in that action. To do so meant that Earth First!ers had to enter into negotiations with the Humboldt County sheriff. They also had to negotiate with Maxxam since there was nowhere for the rally to take place near the gate to Headwaters that was not property of Maxxam. Unless Maxxam agreed to allow the rally, all of the participants would be at risk of arrest for trespass. The negotiations raised concerns for some Earth First! activists who felt that these negotiations violated the no compromise stance of Earth First![21], but Earth First!er Judi Bari played a leading role in the negotiations, and Darryl Cherney strongly defended this approach.[22]

The more inclusive approach resulted in the largest turnout at a Headwaters event since Redwood Summer, with more than 2,000 people present on September 15th. What was even more notable was the level of involvement in the subsequent direct action. Following a structure negotiated with the police, rally participants who wanted to take part in civil disobedience marched to the gate leading to Headwaters and then, one-by-one, walked across the Maxxam property line where police were waiting to arrest them. The event went smoothly. The protestors were taken by buses away from the gate, cited for trespassing, and then released. The September 15th civil disobedience drew a much more diverse group of arrestees than previous Earth First! actions. And the total of 264 arrestees was four times larger than the previous Earth First! record.[23] On the day of the protest, the California Senate voted overwhelmingly to call for negotiations to purchase Headwaters Grove. And before the day was out, EPIC got word that a federal court had granted them a temporary restraining order halting Maxxam's salvage logging. The logging of Headwaters had once again been averted.

The Headwaters Forest Coordinating Committee

The success of the September 15th rally was not simply the result of Earth First!, but also the expanding circle of grassroots Headwaters groups. With the increased threat to Headwaters, the level of coordination among these groups grew. Another impetus for greater coordination was the perceived political opportunity from the upcoming 1996 presidential election. President Clinton faced growing discontent from the environmentalists that he had courted during the 1992 election. The election of Clinton had resulted in little tangible increase in environmental protection. Moreover, in 1995, Clinton had signed a bill containing supplemental language that created a temporary waiver of environmental laws for salvage logging on national forests. (This federal provision was different from the state-level salvage exemption being used by Maxxam.) As described in chapter 4, this "Salvage Rider" led to a surge in logging of old-growth areas in the Pacific Northwest. In light of these events, Clinton needed to woo back environmentalist voters in anticipation of the 1996 presidential election. One possible avenue for the Clinton administration to do so would be to intervene in the Headwaters issue.

Environmental grantmakers became energized over this prospect. By the early 1990s, some Headwaters groups had begun to receive support from a small pool of funders, such as the Foundation for Deep Ecology and the Columbia Foundation, but overall, the level of foundation funding had not matched the level of public interest in Headwaters. However, in 1995, the prospect of federal intervention triggered a surge in funder interest. Tom Van Dyck, a leading figure in socially responsible investing and the founder of the As You Sow Foundation, donated $100,000 to the Headwaters campaign and sought to get other funders to match his grant. However, he found that other grantmakers were reluctant to choose how to distribute funding among the numerous grassroots groups working on this issue. Instead, many foundations preferred to have a single entity that they could give a grant to on this issue. This funder concern became a primary impetus for the Headwaters groups to shift from informal coordination to creating a formal coalition organization through the Headwaters

Forest Coordinating Committee (HFCC). While the groups had worked together extensively up to this time, Van Dyck described this fundraising issue as "the cement that brought the groups together" as a formal entity. Under this new approach, the HFCC would serve as the fundraising body for the grassroots groups in Headwaters campaign, with the Trees Foundation as its fiscal sponsor.

The HFCC organizational structure offered three main advantages to the Headwaters groups. First, it dramatically increased their funding. Within a year, the HFCC received almost $1 million. For groups that had worked for years with virtually no budget, this was an extraordinary sum. For example, the BACH was able to pay some of its organizers for the first time, and EPIC eventually expanded to a paid staff of more than a dozen people.

Second, the HFCC allowed the Headwaters activists to speak with a unified voice on key policy issues. This arrangement became especially important with the prospect of federal acquisition of Headwaters. A concern among many Headwaters activists was that the Clinton administration would only seek to protect the 3,000-acre Headwaters Grove, since it was the most well-known segment of Headwaters Forest. However, from an ecological perspective, protection of the other old-growth groves was also essential for saving the marbled murrelet from extinction. Moreover, a lesson from previous redwood campaigns was that if only the old-growth groves were protected, they would deteriorate. When the forest around those groves was cleared, the trees within the protected area were harmed by that loss of that buffer. Large trees would be toppled by winds blowing through the logged areas. The forest would become hotter and drier, and therefore unfavorable to redwoods. Avian predators of the murrelets that would normally be deterred by the dense old-growth canopy could access the old-growth stands through the surrounding clearcuts. And murrelets would also be more vulnerable to predation if they tried to fly through the logged areas between the stands. In short, without a sufficient forested buffer area beyond the old-growth groves, there would eventually be significant loss of trees and wildlife within the ostensibly protected areas.

For the forest to be viable in the long term, the protected area needed to be significantly larger than just the Headwaters Grove. The Headwaters groups sought to speak with a unified voice on this point. Otherwise, there was concern that the administration might co-opt some groups into endorsing and legitimating a proposal that the rest of the groups saw as unacceptably small. The HFCC operated on consensus and, after extensive discussion, all of the groups agreed to call for the protection of not only Headwaters Grove, but also the five other large old-growth groves in Headwaters Forest, as well as the buffer zone between them, for a total of 60,000 acres.

The third and perhaps most notable benefit of the HFCC was that it offered a grassroots-based alternative to the national environmental organizations. For the most part, the large environmental organizations had not played a major role in the Headwaters campaign. Sierra Club was the most notable exception, though, as discussed earlier, it was primarily the chapter-level volunteers who were most active on this issue. Sierra Club national staff, like other national organizations, would only get involved when a political resolution of the issue was looming, such as with the Hamburg bill. Since the demise of the Hamburg bill, the nationals had largely abandoned the issue. The structure of the HFCC allowed the various small groups to magnify their influence so as to fill the role usually played by the national environmental organizations, while at the same time taking much stronger positions than the nationals would have been likely to assert. As Karen Pickett of BACH summarized, "Except for the Sierra Club, we were all very small organizations with no political clout at all, and yet when we came together, we became something more powerful and we became a force to be reckoned with."

The September 15th Rally of 1996 and the Deal

By summer 1996, the Headwaters Forest Coordinating Committee was engaged in an unprecedented level of Headwaters activism. By then its members were aware of closed-door negotiations between the Clinton administration and Hurwitz. Senator Dianne Feinstein had also gotten involved in these negotiations, which made some of the activists nervous because Feinstein was considered a probusiness Democrat. Meanwhile, Maxxam had found a way around EPIC's injunction on salvage logging, so the prospect of logging in Headwaters once again loomed after the end of murrelet nesting season on September 15th.

The HFCC's member groups spent the summer preparing for another mass mobilization on September 15th. Like the previous year, there would be a rally in Carlotta organized by the HFCC that would conclude with a prearranged symbolic mass trespass at the gate to Headwaters organized by Earth First!. As the rally date approached, Headwaters received extensive news coverage not only in the San Francisco papers but also nationally. The activists had succeeded in making Headwaters a national issue. And for many of these activists, the 1996 rally would be a high point of the campaign.

One day before the rally, Senator Feinstein went on television and radio to announce a vaguely worded, two-week moratorium on salvage operations in Headwaters Grove and to discourage people from attending the rally. Nonetheless, the turnout on September 15th far exceeded organizers' expectations and was unprecedented for the region. About 8,000 people attended the rally and then 1,033 people crossed the Maxxam property line at the gate to Headwaters.[24] That was the largest number of people ever arrested for a forest protection protest in U.S. history.[25] Building on the success of the previous September's orchestrated arrest, the civil disobedience attracted the participation of many people from a variety of ages and backgrounds who might not otherwise take part in direct action. Musician Bonnie Raitt, former congressman Dan Hamburg, and EPIC's Cecelia Lanman were among those who crossed the line. For them, as for many of the participants, it was the first time they had ever been arrested for civil disobedience. The number of arrestees was so large that the police ran out of plastic handcuffs and arrest forms.[26]

On September 16, Earth First! began a series of coordinated blockades to halt the logging in areas near Headwaters Grove. The Earth First!ers used sophisticated techniques to chain themselves to barricades in the logging road in such a way that they could not be removed from the road without first extracting them from the "lock-down." In the past, protestors had used bicycle u-locks to chain their necks to road gates or logging trucks, but police learned to carry a set of u-lock keys, which diminished the effectiveness of this technique. By 1996, Earth First!ers were using much more complex "lock boxes," in which an activist would have her arms chained inside a thick metal tube that the authorities would have to cut through in order to unchain the protester. These lock boxes were sometimes encased in further barriers, such as a concrete-filled barrel or even more complex configurations inserted within the frame of an old car. In each case, the goal was make the extraction of the protester difficult and time consuming, to keep the blockade going as long as possible.

There were also tactical innovations in tree-sitting. No longer satisfied with the simple one-person platform, in fall of 1996, Earth First!ers created one of the most elaborate tree-sits ever concocted to block logging near Owl Creek. It consisted of a 400 square-foot fishing net hung high in the air between redwoods, serving as a hammock connecting the tree-sit platforms on six different trees, such that cutting down any one of the trees would imperil all of the other tree-sitters. This configuration was nicknamed the "Ewok Village" after the forest-dwelling creatures of the Star Wars movies. It hindered logging for nineteen days but was ultimately dismantled by the police and Maxxam tree climbers.[27] In general, the innovative blockade tactics were effective in delaying the logging, but they rarely stopped it entirely.

Then on September 28, Senator Feinstein announced that the government had reached a deal with Maxxam CEO Charles Hurwitz over Headwaters. However, "the deal," as it came to be known, was very troubling to Headwaters activists. The federal government would pay $250 million, and the state of California would pay $130 million, for a total of $380 million going to purchase parts of Headwaters. In exchange, they would receive Headwaters Grove and one smaller adjacent old-growth grove, Elkhead Springs, and a buffer zone around them. The total area would be about 7,500 acres, only about half of which was old growth. Disappointingly, this area was about one-eighth the size of the HFCC proposal and did not include four of the six main old-growth groves (though Headwater Grove was much larger than the other groves). Headwaters activists also expressed concern over Hurwitz receiving such a large sum of money for an area that they believed he could have never cut due to the marbled murrelet's habitat protection under the Endangered Species Act.

The deal was contingent upon Maxxam receiving approval from the federal government for a habitat conservation plan (HCP) that the corporation would create for all of its remaining lands. HCPs were established under an amendment to the Endangered Species Act in 1982. The provision was intended to provide more flexibility in the implementation of the act by allowing for a process whereby landowners could be permitted to destroy endangered species habitat and kill endangered species if they mitigated this damage by creating or improving endangered species habitat elsewhere. For the first decade after it was introduced, the HCP process was rarely used, but under the Clinton administration, the number of HCPs skyrocketed. They were increasingly being offered to whole corporate landholdings rather than individual projects. The mitigations offered by these HCPs were often outweighed by the losses they permitted. And a new "No Surprises" policy meant that once an HCP was authorized, the landowner could be exempted from additional responsibilities for up to 100 years if new endangered species were found on the property or if existing endangered species continued to decline. In short, environmentalists saw the use of HCPs as a loophole for weakening the enforcement of the Endangered Species Act. "It sounds good on paper, but the way that (habitat plans) have worked in reality is that they are very sweet deals for landowners," said Tara Mueller, an environmental lawyer active in the Headwaters campaign.[28] Headwaters activists were concerned that the HCP would provide far less protection for the marbled murrelet habitat on Maxxam's land than if the federal government simply enforced the Endangered Species Act.

Under the deal, Maxxam would also develop a state-level sustained yield plan, whereby it would map out its long-term logging plans in exchange for easing of state regulatory oversight of that logging. The prospect of the HCP along with the sustained yield plan meant that it would become much harder for environmentalists to challenge Maxxam's logging in court. Moreover, as part of the deal, the federal and state agencies would sign a "mutual aid pact" with Maxxam to defend the logging permits resulting from this process from any future litigation. As Kathy Bailey explained, "If they hadn't had that mutual aid pact, we could have filed [litigation] against the feds and state saying that their permits were illegal, and they could have negotiated with us and maybe made some modifications . . . but they were locked into this position of having to be diametrically opposed to us."

Finally, as part of the deal, Hurwitz agreed to drop his takings lawsuit. Most Headwaters activists and lawyers saw this suit as a red herring that lacked grounds for success in the courts, but the Clinton administration had put a strong emphasis on averting the takings suit. To many Headwaters activists, this stance was indicative of the Clinton administration's reluctance to enforce the Endangered Species Act when it conflicted with corporate interests.

In all, the proposed deal was a great disappointment to most Headwaters activists. However, when it was announced by Feinstein in late September 1996, the deal was just an agreement in principle rather than a completed process. The process of getting the funds approved by the federal and state legislatures and the procedures for approving the HCP ensured that it would take at least a couple years before the deal could be finalized. During that time, the question of how Headwaters activists should respond to the prospect of the deal would become a contentious issue within the HFCC.

Funding Issues

The announcement of the deal exacerbated tensions and divisions that were already developing within the Headwaters Forest Coordinating Committee. Those internal rifts would become a drain on the energy of the Headwaters organizers at a time when the Headwaters campaign faced its greatest external challenges. The HFCC brought together a network of activists who had already worked with each other extensively and shared strong feelings of camaraderie. Yet the HFCC ultimately fostered such divisive conflicts that many of the participants still found it difficult to talk about them in interviews conducted almost a decade later. In one of these interviews, Michael Passoff attributed the conflicts to the "strong personalities" involved in the HFCC. However, these strong personalities had been working together effectively for years prior to the formalization of the coalition. Therefore, it is useful to consider the ways that the HFCC's structure contributed to these conflicts.

In the interviews, the Headwaters organizers often distinguished between the benefits of informal coordination among the groups and the problems stemming from the formal structure of the HFCC that was established in order to fit the preferences of grantmaking foundations. For example, Karen Pickett said,

> All the trouble that came from the money part of it, you can kind of separate that part out. That really did come from the funders. They wanted us to be more coordinated in that they wanted to get a [single] grant [proposal] for the Headwaters campaign and for it to be a shared grant. [audible sigh] We did that. We responded to their request, when it would have been more appropriate to say, "That's absolutely ridiculous and impossible." When you look at all of these organizations that have their own infrastructure and decision-making processes, it was such a varied terrain in terms of the funding needs. And it ended up imploding.

The joint fundraising structure of the HFCC was a primary source of the conflicts within the coalition. As Tracy Katelman of the Trees Foundation summarized, "Joint fundraising killed us."[29] As noted earlier, a primary impetus for the creation of the HFCC as a formal entity was that many funders did not want to have to choose how to distribute their grants among a variety of groups playing different roles within the same campaign. Instead, they wanted to give one grant to one entity. The HFCC was then given the responsibility for distributing those funds among its member groups. Those groups were thus financially dependent on the actions and decisions made by the HFCC. Because the member groups all relied on the same pool of funding, if the actions of one group risked alienating funders and losing funding, it could potentially reduce the total amount of funds available to the other groups as well. The stakes were raised on any decision because each decision could now have direct implications for the fiscal well-being of every member group. As a result, decision making became a much more contentious process. Also, under the joint fundraising arrangement, all of the groups in the HFCC had to approve each other's work plans. The time and energy of the Headwaters activists was increasingly redirected into the organizational demands of the HFCC. As Tim Little of the Rose Foundation observed, "Everyone got knocked off of doing what they did well because they were busy supervising everyone else."[30] Previously the diversity of tactics and approaches among the Headwaters groups had been a source of strength for the campaign. However, when the different groups felt compelled to manage each other within the joint fundraising structure, those differences were more likely to become a source of conflict as their divergent and sometimes clashing approaches vied to steer the direction of the HFCC. Under the HFCC structure, the diversity among Headwaters groups changed from one of the campaign's greatest strengths into a liability.

Another consequence of the joint fundraising structure was the hiring of an outside professional to serve as the coordinator for the HFCC. Pickett noted, "That was a requirement of the funders. If you are going to do joint fundraising and you have eight organizations, then [the funders] want to talk to one person. Suddenly, you're supposed to create this artificial infrastructure that doesn't really benefit the strategy of the campaign, and probably actually takes away from the actual work." The hiring of Doug Linney—a former political director for the California League of Conservation Voters—to be the HFCC's coordinator was a particular source of consternation for many of the longtime Headwaters organizers. Tracey

Katelman identified the hiring of Linney as "part of where [the HFCC] started to fall apart."

The criticisms of Linney had less to do with him personally than with the grassroots activists' general concern over the growing role of professional consultants in the Headwaters campaign. This development was encouraged by the increased funding for the HFCC by donors who favored greater use of consultants from outside the campaign. Cherney provided an ironic description of how this has been a recurrent pattern when grassroots campaigns get sizable outside funding:

> There's something about money where people say you have to hire a consultant: "Now that we have money, we're going to hire the pros. We've done this really, really great job, [but] we're not going to pay [The Man Who Walks in the] Woods, we're not going to pay Judi [Bari], we're not going to pay Darryl [Cherney], we're not going to pay Karen [Pickett]. We're just going to hire strangers because strangers can do it better than we can" [said with sarcasm]. That is so common. This is not unique to us at all. And ultimately if those people do not live up to expectations—and they inevitably will not—it creates internal fighting, bickering, and lackluster results.

The grassroots activists had a difficult time trusting the outside professionals hired to work with the HFCC. They were thought to be more likely to see the campaign as a temporary job rather than a long-term commitment. As Cherney noted, "once the money was gone, [they] were on to the next issue." The professional consultants were also associated with insider political strategies, which meant they were more open to compromising at a time when compromise around the deal was an especially sensitive issue.

Concern with outside professionals was not limited to those hired by the HFCC. With increased funding available for Headwaters work, new organizations suddenly took interest in getting grants to work on this issue. In particular, many of the participants in the campaign spoke critically of an organization formed during this time called the Headwaters Sanctuary Project. This entity was created by Ted Nordhaus, a political consultant who would later coauthor the essay "The Death of Environmentalism."[31] Cherney dubbed that organization the "Headwaters Mercenary Project" after Nordhaus received a grant that the HFCC groups believed was originally intended for them. This experience exacerbated the growing distrust of outside professionals getting involved in Headwaters.

This mistrust marked a shift in the dynamics of the campaign. For its first decade, the Headwaters campaign had operated on very limited resources and most of the activists involved were not paid to work on the issue. That meant that the participants in that campaign were involved first and foremost because of their concern over the fate of Headwaters Forest rather than any sort of monetary incentives. While the various groups might have different tactics or styles, they were united by a bond of trust in each other based in the knowledge that everyone was participating because of a shared commitment to saving the forest. This provided a foundation for effective cooperation among a diverse set of groups. However, as the funding for Headwaters advocacy grew dramatically, longtime Headwaters activists could no longer be confident that newer, paid participants were motivated primarily by concern for the forests. The trust that had been the basis for coordination thus diminished amid the HFCC's fundraising success.[32]

In light of these problems, it is not surprising that Headwaters activists were ambivalent about the impact that a large infusion of money had on their campaign. From the more professional end of the spectrum, Tom Van Dyck, as a major donor and fundraiser for the HFCC, assumed that the *lack* of resources was the source of tension for the small groups and therefore that increased funding would reduce the conflicts within the campaign. In contrast, many of the grassroots activists deemphasized the importance of the money and even saw it as a source of problems. Tracy Katelman said, "We don't need the money as much as we need the spirit. What kept the campaign going up here really was that spirit. Some of that got compromised for money and that's where people started getting burned out." Cherney even argued that "we could have done most of what we did without that money." Between the poles represented by Van Dyck and Cherney, Pickett summarized the contradictory effects of money within the Headwaters campaign: "The infusion of money into this tight, dynamic, effective group poked holes in it. But we needed the money, so it was really a catch-22."

Divisions over the Deal

The problems faced by the Headwaters Forest Coordinating Committee cannot be attributed solely to funders and funding. The prospect of responding to the Headwaters deal created difficulties for the decision-making system of the HFCC. The HFCC operated on consensus, which meant that all of the member groups had to support a decision before it would be adopted. The challenges of reaching consensus led to long, grueling meetings, but they also allowed all of the groups to be heard and feel invested in the outcome. At the beginning, consensus had worked well for the HFCC in formulating a unified demand for the protection of six old-growth groves and 60,000 acres of Headwaters Forest. However when the tentative deal was announced by Feinstein, the groups were faced with the possibility of an outcome that fell far short of their demands, involving only two groves and 7,500 acres, as well as the dangers from the habitat conservation plan and the bad precedent of a large payment to Hurwitz. The prospective deal drove a wedge into the Headwaters coalition. All of the HFCC groups agreed that the deal was inadequate and flawed, but serious divisions emerged over the extent to which they were willing to criticize the deal. The consensus decision-making system was strained by those divisions.

One source of division was the differing levels of concern over the impact that criticism of the deal could have on Democratic politicians such as Senator Feinstein and President Clinton who had brokered the deal. Both the Sierra Club and the Rose Foundation had a stake in maintaining good relations with the Democratic politicians. Jill Ratner of the Rose Foundation was deeply involved in support for the Democratic Party. For Ratner, the deal had implications not only for Headwaters Forest, but also for the long-term prospects of Democratic politicians that she saw as allies: "[The Rose Foundation] might have cared more in the long run whether what we did helped or hurt certain people in government, whereas I think that was pretty irrelevant to most of the [Headwaters groups] in Humboldt."

Those other groups were more than willing to publicly criticize the Democratic officials who they saw leading the deal-making process. This was expressed most dramatically soon after the deal was announced when Earth First! staged a protest at the Democratic Party headquarters in Humboldt. To illustrate their theme that the deal was "bullshit and fluff," some Earth First!ers dumped manure and feathers in the office. (At the end of the protest, the Earth First!ers rented an industrial vacuum cleaner and cleaned up the office. One participant reported, "We all agreed that we didn't have a problem with the local workers. I think it was cleaner than when we found it."[33]) Sierra Club Executive Director Carl Pope publicly condemned the Earth First! action in a San Francisco newspaper, a move that served as an early marker of the cracks forming in the Headwaters coalition.[34]

While later actions were not as provocative, grassroots Headwaters groups continued to make Feinstein and Clinton primary targets for their criticisms of the deal. For example, at some protests, a member of BACH would dress as Feinstein and lie in a motorized moving bed with a Charles Hurwitz impersonator and piles of money. These activists saw Feinstein and Clinton as primarily interested in getting credit for an environmental victory in order to boost their stature with voters before the November election, regardless of whether that "victory" truly protected Headwaters Forest or not. From this perspective, the Headwaters campaign needed to debunk the image of the deal as being an environmental victory in order to create pressure for stronger protection. Feinstein and Clinton were central architects of the deal and thus criticism of it required criticism of Feinstein and Clinton. Moreover, environmental organizations unwilling to strongly criticize Feinstein and Clinton were seen to be softening their criticisms of the deal and putting the interests of the Democratic Party before the needs of the forest. Thus, groups with close ties to the Democratic Party were viewed with suspicion by the other Headwaters activists in the wake of the deal's announcement.

Beyond differences over their relationships to the Democratic Party, a primary division among the Headwaters groups centered around debates over whether extensive criticism of the deal might ultimately derail it. The position of most of the HFCC groups was summed up by Cecelia Lanman: "Better no deal than a bad deal." These groups took a principled perspective, contending that they should serve as unconstrained advocates for the needs of the forest ecosystem. For them, the deal needed to be strongly criticized and, if it remained deeply flawed, then it should be stopped. However, a few groups—particularly the Sierra Club and the Rose Foundation—were reluctant to risk having the deal fail. They took a more pragmatic approach. They thought that the campaign should take what was attainable under the existing political conditions, as represented by the deal. Once the main grove was protected, they believed that the campaign could then turn its attention to protecting more of the forest. In defending this approach, those coming from the pragmatic perspective emphasized their fears that, without the deal, Headwaters Grove might be logged.

By contrast, most other Headwaters groups emphasized concerns over the ways that the deal would harm the protection of the larger Headwaters Forest. Preservation of the Headwaters Grove could foster the misperception that Headwaters as a whole had been saved and thus have a strong demobilizing effect for the campaign. Moreover, the habitat conservation plan would take away much of the basis for legal challenges to protect the rest of Headwaters. And the large payment to Hurwitz would not only set a bad precedent, but also make him a more powerful opponent.

These divergent perceptions of the deal are best understood within the context of a split among the Headwaters groups between insider and outsider strategies. Unlike most of the HFCC groups, the Sierra Club and Rose Foundation relied more on an insider strategy, based on access to the elected officials involved in crafting the deal. That strategy encouraged a more pragmatic approach, as Jill Ratner of the Rose Foundation explained:

> At first we tried to be a neutral bridge [to the rest of the HFCC], but as time goes on, I became invested in things that I saw as opportunities or achievable results. And so that would create more tension because there is a tendency for insiders to focus on what can we get done here and outsiders to be more focused on what are the needs of the community or what are the ultimate stakes. I think that is a tension, and a tough one to avoid really. . . . To the extent that I screwed up in the way I did my work on Headwaters, it was because I was too limited by what I thought was attainable in thinking about what was necessary. And I think that is a lesson. You do honestly need to keep your eye on the prize. It's also true that if you only keep your eye on the prize, you will never get anything.

The insider approach was based on a model of social change that was ultimately rooted in conventional political processes. The insider groups saw their power in their access to and influence on elected officials. For them, the deal represented more or less the best that they could get under the political conditions at the time. Moreover, if the deal failed, they saw it as unlikely that the Clinton administration would take up the Headwaters issue again. Thus, for groups that saw social change primarily in terms of the actions of elected officials, there was no clear prospect for federal action to protect Headwaters without the deal. It is not surprising then that groups coming from this perspective would fear that the forest would be cut.

The Headwaters groups relying on an outsider strategy worked from a different underlying model of social change. They saw their power as coming from outside the political sphere.[35] The grassroots groups had successfully used a variety of tactics to block the logging of Headwaters. For example, EPIC used litigation, Earth First! used direct action, and BACH mobilized the urban constituencies for mass civil disobedience. Using these tactics, they had kept Maxxam from logging the old growth in Headwaters Grove since 1989 and significantly hindered logging in the rest of the forest. Through an uncompromising defense of the forest, they had built widespread support for the protection of Headwaters. Thus, there was a perception among these activists that if the deal failed, Maxxam could still be stopped from logging Headwaters Grove. The Headwaters groups could continue to use the courts and the Endangered Species Act to halt the logging. And even if the judiciary failed to stop Maxxam, many Headwaters activists believed that the public simply would not allow Headwaters to be logged; if Maxxam attempted to cut down the main grove, it would set off mass resistance and upheaval. They were not alone in this perception. Senator Feinstein stated, "The alternative to that solution [the deal] is devastating. It would be World War III in the Headwaters Forest."[36] While not blasé about the safety of Headwaters Grove, the grassroots Headwaters activists saw their power to protect it as existing outside of the electoral sphere and thus did not see the need to accept a political compromise that could weaken their ability to save the rest of the forest. As a result, they saw the efforts of other groups to preserve the deal as contrary to the needs of forest.

The issue of how to respond to the deal exposed and exacerbated a sharp rift within the HFCC. While a diversity of tactics had been advantageous to the campaign, the differences over insider versus outsider strategies would prove more detrimental. The HFCC sought to speak with one voice. This task had been comparatively easy in the HFCC's early days when the member groups simply had to agree on what they wanted to ask for in regards to level of protection for the forest. However, after the announcement of the deal, groups faced the much more difficult choices about what level of protection they would ultimately accept. As Ratner summarized, "It is much easier to agree in concept, but once the details were on the table and it was clear that anything we were going to get, we were going to get in this process, then it was much tougher to get along together because the differences seemed to have more impact."

In this context, the strategic differences between insider and outsider groups became a barrier to consensus and thus a problem. Consequently, the HFCC was not able to produce a truly unified response to the proposed deal. Moreover, the challenge of trying to create consensus exacerbated conflicts among the groups that had already been simmering due to the problems stemming from the joint fundraising. The growing animosity around these conflicts caused the Rose Foundation to leave the HFCC, and by spring of 1997 the HFCC was more or less defunct as a formal structure. After the demise of the HFCC, the Headwaters groups resumed their informal coordination without the joint fundraising system, but this shift came too late. The HFCC structure had fostered conflicts between Headwaters groups at exactly the point when a unified response would have been most useful. Moreover, these internal conflicts sapped the energy of the organizers at a time when the external challenges for the campaign were most pronounced.

Organizing Challenges in the Wake of the Deal Announcement

Beyond the internal problems with the HFCC, the Headwaters groups also faced the external challenge of organizing in the wake of the deal announcement. At one level, the various groups continued the work that they had been doing all along. However, this changed context presented at least three major difficulties. For one, organizers had to confront the widespread misperception that all of Headwaters Forest had been saved, which made it much harder for the activists to generate media attention and mobilize the public around the issue.

The groups struggled to keep Headwaters in the public eye, and it was often only the most controversial actions that would receive media coverage. As described earlier, there was the Earth First! action at Senator Feinstein's Humboldt office in late September 1996. Then in late November, Hollywood celebrity Woody Harrelson participated in an action with other Headwaters activists to hang a massive banner from atop the Golden Gate Bridge that read, "Hurwitz, Aren't Redwoods More Precious than Gold?" While the banner hang was successful in garnering news coverage, much of that coverage was rather critical because the action had caused a five-hour shutdown of traffic into San Francisco. By comparison, creative but less contentious actions—such as a Halloween sit-in in Senator

Feinstein's San Francisco office by thirteen protestors wearing Charles Hurwitz masks—were not reported in the major newspapers.

A second challenge was the increasingly technical nature of the discussions of Headwaters protection. Whereas previous public mobilizations had centered on the immediate threat of Headwaters being cut, now the main focus of organizing was the public hearings on the proposed Maxxam habitat conservation plan, a rather technical process. Nonetheless, the Headwaters groups conducted a large-scale campaign to educate the public about the habitat conservation plan process and the preparation of comments for the hearings. Despite the challenges, Headwaters campaigners were very successful in getting mass participation in the process. There were more than 18,000 written comments received on the plan.

A third challenge was the lack of clarity on whether the goal of this organizing was to stop the deal or to reform it. Moreover, if it could be stopped, there was the question of what would take its place. The Trees Foundation prepared a "Headwaters Forest Stewardship Plan" as a grassroots-crafted substitute for the deal. It was extensively researched, included sophisticated computer mapping, and was widely praised by other activists; yet it never became a viable the alternative to the deal.

On top of these challenges, in October 1996, Earth First! organizer Judi Bari was diagnosed with an aggressive form of cancer. She would have only five more months to live. During that time, much of her energy was put into the civil rights lawsuit she and Darryl Cherney had brought against the FBI and Oakland police over their handling of the 1990 bombing. The suit had been filed in 1991, but the FBI's delay tactics prevented a trial date from being set. When it became clear that Bari would not live to see the case come to trial, she made preparations that would allow the suit to continue without her. She worked intensively on the case up until a few days before she died on March 2, 1997. With Bari's passing, the Headwaters campaign lost one of its central organizers. As Jill Ratner remembered, "Her absence created a vacuum that really nobody else could fill."

The effects of this loss were evident as the Headwaters groups prepared for their third annual rally at the end of marbled murrelet nesting season in September 1997. Bari had been a principal negotiator for those earlier two events and her absence was keenly felt in 1997. This time, Maxxam would not allow the rally to take place on its land and the police would not make any commitments regarding their response to civil disobedience. Organizers found an alternate location away from the gate to

Headwaters, and the rally once again attracted thousands of participants. However, concerns over the prospect of police violence caused some of the organizers to cancel the planned civil disobedience component of the event at the last minute, even though a larger group of organizers had earlier reached consensus to proceed with it. The outcome felt like a missed opportunity to many of the participants. And after that, the Headwaters campaign never organized another September 15th rally or any further mass civil disobedience.

Increasing Repression: Pepper Spray and the Death of David "Gypsy" Chain

Despite the absence of a mass civil disobedience at the 1997 rally, Earth First!ers continued to organize direct actions to slow logging near Headwaters. However, they faced the challenge of growing levels of repression and violence from both the police and the loggers. The police response to the Earth First! blockade techniques changed as they grew frustrated with the increasingly complex lockboxes. Despite the name lockbox, a protestor was usually not actually locked within this apparatus; instead she was attached by a clamp attached to a chain around the protestor's wrist so that the she could detach herself from the lockbox at any point if she chose to do so. Some police began opting to inflict pain on the protestors to try to coerce them into detaching themselves rather than spend time cutting through the layers of metal of the lockbox to extract them.

At first this practice occurred away from public sight, but the police tactic drew national attention after Humboldt County sheriff's deputies used pepper spray on Headwaters activists during a sit-in the Eureka office of Congressman Frank Riggs on October 16, 1997. Four women used lockboxes to link their arms in a circle as they sat on the floor of the office. The deputies opted to force the activists to detach themselves by applying pepper spray on q-tips that were swabbed against the women's eyes to produce intense pain. The sheriff's department videotaped the incident, including the pepper spraying. When the protesters obtained a copy of the video two weeks later, they released it to the media. The footage provided vivid images of the protesters' suffering and received extensive coverage in the national media. The protesters also filed a lawsuit against the sheriff over the Riggs office incident, along with two other pepper spray incidents in the previous month, though it would take until 2005 for the plaintiffs to prevail in court.

The violence against Headwaters protesters turned deadly on September 17, 1998, when Earth First!er David "Gypsy" Chain was killed after Pacific Lumber logger A. E. Ammons felled a tree that landed on top of the activist. At the time, Gypsy and eight other Earth First!ers were using the "cat-and-mouse" tactic, wherein activists would enter active logging sites in order to delay the logging. The loggers were supposed to stop cutting when other people were in the vicinity because of the danger from the falling trees, but in this case Ammons did not. Immediately, following Gypsy's death, Maxxam/PL's spokesperson claimed that the logger had been unaware of the presence of the protestors in the area.[37] However, a videotape taken by the Earth First! activists recorded Ammons angrily berating them and shouting, "Get outta here! Otherwise I'll fuckin', I'll make sure I got a tree comin' this way!" shortly before cutting the tree that killed Gypsy.[38]

Despite the video, the Humboldt County district attorney would not file charges against Ammons, so Gypsy's family filed suit against Maxxam and the loggers. When he was killed, Gypsy was in Grizzly Creek area. The action that day had intended to delay the logging while trying to get a California Department of Forestry official to come check reports of illegal logging activity at the site. On the day after Gypsy's death, CDF visited the site and found two violations.[39] One of the other participants in that action wrote, "Gypsy died doing CDF's job."[40]

Gypsy was the first Earth First!er killed doing direct action in the group's eighteen-year history. His death and the pepper spraying of other Headwaters activists served as vivid reminders of the risks that accompanied the choice of tactics of the Earth First!ers in the Headwaters campaign. Having disavowed the monkeywrenching associated with the early years of Earth First!, the Headwaters Earth First!ers relied on a set of tactics that centered around putting their bodies directly at risk in order to stop the logging—for example, lockdowns, cat-and-mouse, and tree-sits. This approach almost guaranteed the arrest of the activists. It also put the activists in potential jeopardy of injury and pain, as in the case of the pepper spray victims, or even death, as with Gypsy. Ultimately, the renunciation of tree-spiking and monkeywrenching allowed Earth First! to build a broader base of support in the community, but it came at a high personal cost to the activists.[41]

New Allies: Salmon Protection, Rural Communities, and Steelworkers

Even amid all of the challenges following the Headwaters deal announcement, the campaign continued to see a proliferation of innovative new tactics to confront Maxxam. Early 1997 saw the launch of the old-growth
redwood boycott by the Rainforest Action Network, as well as the Maxxam
shareholder initiatives of As You Sow and Rose Foundation. Also in spring
1997, the coho salmon was listed as a threatened species from California
to Oregon. These salmon relied on rivers and streams near the coast as their
spawning habitat, including the aptly named Salmon Creek in Headwaters
Forest. Logging led to erosion that increased the amount of sediment in
streams while the loss of shade from tree cover near the streams raised the
water temperature, both of which made conditions less hospitable for spawning salmon. The listing of the coho held out the prospect of Endangered
Species Act restrictions on logging not only of old growth, but of any trees
near salmon watersheds. The threatened status of the coho also illustrated
the harm that the timber industry was causing to the fishing industry, which
faced declining salmon numbers as a result of the effects of logging.

Another group harmed by logging was the local residents of Humboldt
County living near Maxxam's clearcuts. Clearcutting on steep slopes
removed the trees that helped keep the soil in place. With the trees gone,
the frequent rainstorms in the area could dislodge the soil, leading to landslides or clogging streams and causing flooding. This point was vividly illustrated when landslides emanating from a Maxxam clearcut destroyed seven
houses in the small town of Stafford on New Year's Eve, 1996. Stafford
was not a counterculture town like Garberville. Its residents had been generally supportive of the timber industry, but Maxxam's tepid response to
the damage from the landslide angered Stafford residents, who filed a lawsuit against the company. One of the Stafford residents, Mike O'Neal, also
allowed his land to be used as the site of the September 1997 Headwaters
rally, and rally participants helped build a sandbag wall to protect the property from further landslides. These developments encouraged Headwaters
groups such as the Humboldt Watershed Council to increase their outreach to the rural communities impacted by Maxxam's logging.

The most well-known outgrowth of the Stafford slide was Julia "Butterfly" Hill's tree-sit in an old-growth redwood dubbed Luna. Luna was located
within another Maxxam timber harvest plan above Stafford immediately

adjacent to where the New Year's landslide had begun. Julia Butterfly ultimately stayed in her tree-sit for an unprecedented two years and helped generate extensive media coverage for the plight of the redwoods (as I discuss later).

Another important development in the Headwaters campaign was the emergence of an alliance between Earth First! and the United Steelworkers of America. This was the union for the striking workers of Kaiser Aluminum. Kaiser, like Pacific Lumber, had been a family-owned company with a reputation for taking good care of its workers until it was taken over by Maxxam. Maxxam sought to force extensive concessions from the Kaiser workers, resulting in a long strike that began in fall 1998. Maxxam then began encouraging laid-off timber workers from Pacific Lumber to work as strikebreakers at Kaiser plants. The steelworkers union tried to build connections to PL workers in order to prevent this flow of strikebreakers, but PL workers were not unionized and the steelworkers' overtures were largely ignored. However some steelworkers discovered Darryl Cherney's "Jail Hurwitz" web site and contacted him. The Earth First!ers were particularly aware of the potential importance of a labor/environmental alliance, given the legacy of Judi Bari's pioneering work in this area. Earth First!er Mike Jakubal volunteered to help the steelworkers union by doing undercover research as a strikebreaker at one of the Kaiser plants.[42] Cherney joined the steelworkers' picket lines and performed his humorous songs lampooning Maxxam and Charles Hurwitz.

Seeing common cause in their opposition to Hurwitz, the alliance between the Headwaters activists and the steelworkers blossomed. The steelworkers learned about the Headwaters deal and saw the potential dangers to their strike if Maxxam were to receive an infusion of hundreds of millions of dollars from it. In November 1998, some steelworkers traveled to the Headwaters habitat conservation plan hearings to voice their criticisms of the deal. Before the habitat conservation plan comment period ended, steelworker union members submitted a thousand comments opposing it.[43] The steelworkers and Headwaters activists held rallies together throughout Humboldt County and the Bay Area. Steelworkers visited Julia Butterfly in her tree-sit. And in December 1998, Earth First!ers and labor activists organized an action at a Washington port to block ships from bringing ore for Kaiser's strikebreaker-operated facilities. At the protest, activists hung a banner that declared, "Charles Hurwitz Cuts Jobs Like He Cuts Trees."[44]

The Culmination of the Deal

Despite the Headwaters activists' protests, the process for finalizing the deal proceeded onward. In November 1997, President Clinton signed an appropriations bill for the Department of the Interior that included $250 million for the acquisition of Headwaters. The approval of the federal funding did not seem to be notably influenced by the concerns of the environmentalists. However, the culmination of the deal also depended on the California state government contributing $130 million to complete the $380 million price tag. There were many supporters of environmental protection in the California State Assembly. At the same time, many among them were reluctant to appropriate substantial funding for a controversial figure such as Charles Hurwitz. Therefore, Headwaters groups found more leverage to challenge the deal at the state level.

Opponents of the deal had an ally in Tom Hayden, who previously helped grassroots activists oppose the Grand Accord as a state assemblyman and was subsequently elected to California's state senate. Hayden allowed members of EPIC and Earth First! to use his office during their lobbying visits to Sacramento to critique the deal. Meanwhile, although the Sierra Club remained critical of the deal, it worked closely with state Senator Byron Sher on a bill to authorize the state's portion of the funding, while at the same time seeking to increase the protections provided under the deal. The Sierra Club's greatest leverage was the connection between California Senate president John Burton and Sierra Club executive director Carl Pope. Burton made it clear that, as president of the senate, he would not allow Headwaters funding legislation to move forward unless it was deemed adequate by Pope. Pope in turn relied primarily on Kathy Bailey to make that assessment. Bailey noted, "We could have stopped that legislation. If Sierra Club had said, 'Pull the plug,' Burton would have pulled the plug. We would have not had the state funding and the whole thing would have collapsed."

However, rather than kill the bill, Bailey used the Sierra Club's leverage to negotiate increased protection of the other old-growth groves not acquired by the deal. Through these negotiations, fifty-year restrictions on logging were added to the other four Headwaters groves as well as some smaller stands of old growth. After Maxxam pushed to be able to log either the remaining parts of Owl Creek or a stand of old growth near Grizzly Creek, the state agreed to spend an additional $100 million to purchase both of those groves. In addition, concerns over the protection of habitat for the newly listed coho salmon led to a requirement in the state legislation for at least 30-foot-wide no-logging buffers along fish-bearing streams.

Not all Sierra Club activists wanted the deal to go forward. Bailey's chief opponent within the Club was Josh Kaufman, a former EPIC activist who had become the conservation chair for the Redwood Chapter of the Sierra Club in northern California. Kaufman led the chapter in outspoken opposition to the deal. He also engaged in outreach to other Sierra Club chapters in order to prevent the Club from supporting the state legislation. However, he was outmaneuvered by Bailey. Kaufman explained, "I think her view prevailed to a large extent because her view was consistent with the way the Club likes to operate—how the Club's staff and the whole Club bureaucracy likes to operate—which is stages of compromises in whatever the big issues of the day are." Kaufman was particularly concerned by the process through which these decisions were made within the Club. He noted, "That was my problem with the Sierra Club during the whole process. Ultimately, Kathy and the staff were making decisions unilaterally and there really wasn't much of a democratic process going on."

From Kathy Bailey's perspective, her actions were supported by other activists whose groups were trapped by their radical rhetoric in opposing the deal, but who still saw that deal as the best hope for protecting Headwaters:

> Despite the fact that I was doing the politically incorrect thing by saying that we should go forward, I felt that the majority of the people I cared about most—in fact everyone I cared about most—really in their heart of hearts did want the thing to go forward. Maybe that's a misinterpretation on my part. I don't think so. So I forged ahead.

However, this silent majority was not evident in the interviews with other Headwaters activists. Indeed, Karen Pickett directly rejected it— "It wasn't like the Sierra Club was being the front for the grassroots groups who didn't want to compromise, but who secretly wanted it to go through because the stakes were so high."

As the legislative session wound to a close, Headwaters activists outside of the Sierra Club such as Cecelia Lanman and Darryl Cherney lobbied to stop the deal. Meanwhile, Bailey was working hard to keep prologging loopholes out of state Senator Sher's legislation to amend the funding of the deal. Relations between Bailey and the other Headwaters activists were cordial, but the division was still evident. At one point Lanman was forcibly removed from Sher's office, and at another time Bailey was asked to leave Hayden's office. On August 31, 1998—the final day of the legislative session—Bailey conveyed her approval of the Sher bill language to Carl Pope, and Pope called Senate President Burton to move the legislation forward. As the state senate approved the bill, the normally soft-spoken Lanman shouted "This is extortion!" from the viewing area, and she was removed from the Capitol. The assembly version of the bill also passed that night, though the official clock had to be stopped when it became apparent that the assembly would not be able to pass the legislation by the midnight deadline. But after midnight enough support was secured, and the assembly bill passed by one vote.

The legislation was then sent on to outgoing Governor Pete Wilson, who signed the state's portion of the Headwaters deal on September 19, just two days after Earth First!er David "Gyspy" Chain was killed while protesting Maxxam logging violations near Grizzly Creek. By November 1998, Maxxam/Pacific Lumber had accumulated about 300 violations of state forestry laws over the previous three years and the California Department of Forestry suspended its timber operator's license. This was the first time in California that a major timber company had lost its license. Maxxam could still cut its trees, but to do so it now had to use subcontracted loggers with their own licenses rather than its employees.

A few days after the news of the license suspension, Sierra Club Executive Director Carl Pope had a cappuccino cream pie thrown in his face by a member of the Biotic Baking Brigade, who declared "This is for Headwaters, Carl!"[45] The Biotic Baking Brigade was an ad hoc Earth First! offshoot of pranksters who threw pies at those they saw as prominent environmental wrongdoers in order to satirize and embarrass them. The Biotic Baking Brigade had first gained notoriety for pie-ing Charles Hurwitz of Maxxam.[46] The group's targeting of Pope was indicative of the frustration among many grassroots environmentalists over the role of the Sierra Club in the passage of the Headwaters deal.

Meanwhile, intensive negotiations continued over provisions in the final versions of the Headwaters habitat conservation plan and the sustained yield plan. The deal needed to be finalized by March 1, 1999 or the federal funding would be revoked. But then on Friday, February 26, Hurwitz announced that he was rejecting the deal. That led to a frantic weekend round of negotiations, resulting in an increase in Pacific Lumber's annual allowable cut by over 30 percent.[47] Headwaters activists expressed doubts that Hurwitz ever truly intended to kill the deal. Moreover, on Monday, Maxxam's stock dropped by 18 percent on news of Hurwitz's actions.[48] Regardless of his motivations, Hurwitz agreed to the revised deal, including the increased logging rates, at two minutes before the federal funding was set to expire.

After more than a decade of struggle, Headwaters Grove was secured under government protection. Yet because that protection was tied to the other facets of the deal, the response among most Headwaters activists was subdued. As Darryl Cherney noted, "There was no party that we held to celebrate the passage of the Headwaters Forest Reserve." But then he added, "I think Ted Nordhaus held a party." Nordhaus—a relative newcomer to the campaign, whose organization was referred to as the Headwaters Mercenary Project by longtime activists—was the only figure directly associated with the Headwaters campaign to offer an upbeat assessment of the finalized deal to the media.[49] Beyond Nordhaus, reporters had to go to the staff of The Wilderness Society—an organization that had not played a major role in the campaign—in order to get quotes celebrating the deal.[50]

Was the Headwaters Deal a Victory?

In interviews conducted a half decade after the finalization of the deal, there continued to be divergent perspectives on it among Headwaters activists, and their positions were largely unchanged since the time of the state funding vote. Kathy Bailey remained the most articulate defender of the deal, even while she was also critical of it: "As much as I am beside myself with what the company is doing under the HCP—not that it was a surprise, though actually they are managing to do worse than even I anticipated—I still don't regret doing it."

By comparison, most of the other Headwaters activists interviewed had a more negative view of the deal. Some were ambivalent about it—on the one hand, they were relieved that Headwaters Grove was securely protected, but also they were quite critical of the problems with the deal. The strongest criticisms of it were expressed by activists who were directly involved in the postdeal forest struggles with Maxxam, such as Cynthia Elkins, a prominent EPIC staff member in the period after Cecelia Lanman stepped down as executive director in 1999. Elkins said, "[The deal] was definitely not a good one for the public or the workers or the environment. Certainly since the deal has come down it has made things even more challenging in a lot of ways." Likewise, Josh Kaufman concluded, "I don't think anything good came out of it." Those activists contended that they would have rather seen the deal stopped.

The starkest criticism of the deal came from Ken Miller of the Humboldt Watershed Council, who saw Maxxam as the main beneficiary. He pointed to research that Maxxam had commissioned in 1989 indicating that they would make the biggest profit by selling the main grove and liquidating the rest of the forest. By supporting the deal, Miller said, "the Sierra Club played right into their hands." For Miller, the passage of the deal had the effect of channeling the Headwaters activists' work into the outcome sought by Hurwitz: "I think we were used to publicize Headwaters and to make sure it had public support for spending so much money. I think that we did [Maxxam's] work for them in terms of media and in terms of getting environmental support for a deal that was basically put together and initiated by Hurwitz back as soon as he took over [Pacific Lumber]."

To understand these divergent perspectives on the legacy of the deal, it is helpful to separate out the various components of the final accord: the protections added by the California legislature; the final versions of the federal habitat conservation plan and state-level sustained yield plan; the direct acquisition of Headwaters Grove and an adjacent grove, plus two additional groves purchased by California; the payment to Maxxam; and Maxxam's withdrawal of its property takings claim against the federal government.

The most significant changes to the deal after it was first announced in 1996 were the protections added under the California state-level funding authorization process, due in large part to the work of Kathy Bailey. For Bailey, the fifty-year restrictions on logging in the old-growth groves that had not been purchased by the government were especially important. Without those additions, she said, "this deal would not have been worth doing." Ken Miller, despite his opposition to the deal, acknowledged that the protection for the other old-growth groves and for the stream buffers were one of the few aspects of the deal-making process that took Hurwitz by surprise. However, in general, critics argued that the deal had not increased the de facto protection for Headwaters, while the HCP had actually reduced protection of old-growth trees outside of these groves. As Cynthia Elkins explained,

> Areas that were protected before, that were off-limits to logging— ancient forests—were opened up to logging. And I think that's something that a lot of people don't understand. The deal in essence didn't really protect any areas. It put them into public ownership, but those areas that were purchased were already off-limits to logging because of EPIC's murrelet case. And through the Headwaters deal, other areas that had been closed off to logging for the protection of the murrelet were sacrificed. So we lost ground through the Headwaters deal, and we paid them a lot of money for it.

Responding to criticisms of the habitat conservation plan, both Kathy Bailey and Jill Ratner noted that Maxxam could have gotten an HCP without the Headwaters deal. It is indeed true that the Clinton administration was an enthusiastic promoter of HCPs. However, because the Headwaters HCP was tied to the deal, it developed within a context that was notably different than if Maxxam had sought an HCP independently. The passage of the deal was contingent on the U.S. Fish and Wildlife Service's approval of an HCP for Maxxam. As such, the HCP process was much more politicized. This potentially could have either positive or negative effects on the outcome. On the one hand, it fostered more scrutiny of the development of the Maxxam HCP. Bailey wrote, "The public and media spotlight on the HCP did help generate scientific and legal support for regional wildlife agency personnel attempting to craft a document that met the letter of the law."[51] But there was also strong pressure on agency officials to approve an HCP rather than be responsible for blocking a prominent deal brokered by the Clinton administration. Indeed, Elkins reported that the Fish and Wildlife Service staff had determined that Maxxam's HCP would jeopardize the survival of the marbled murrelet, but that finding was reversed by their superiors. Similar pressure on state agency staff was reported by California Pubic Employees for Environmental Responsibility: "According to state employees, weaknesses in the Headwaters HCP stem from the tremendous political pressure to make the deal, and the fact that [Pacific Lumber] employed an army of attorneys as negotiators, while the state had mid-level managers and biologists at the bargaining table."[52] Above all, the most significant difference between a standalone HCP and an HCP tied to the Headwaters deal was that the latter included a mutual defense pact between the federal and state governments and Maxxam to defend the HCP against any outside legal challenges. The pact made it much more difficult for environmentalists to litigate over the HCP. If Maxxam had instead gotten a standalone HCP, there would have been no mutual defense pact and the Headwaters activists would have been in a far stronger position to challenge any aspect of the HCP that reduced protection for the forests.

Beyond the HCP and the sustained yield plan, the core of the deal was the acquisition of Headwaters Grove and, ultimately, three smaller old-growth stands. In general, the activists interviewed were glad that Headwaters Grove was securely protected under public ownership. However, there was notable disagreement over whether those trees would have been protected from logging even without the deal. Critics of the deal emphasized that legal precedent established by the marbled murrelet case for protecting the old-growth redwoods under the Endangered Species Act. While acknowledging that there was some risk and uncertainty involved, many Headwaters activists felt that some combination of litigation, public outrage, and direct action would likely have protected Headwaters Grove from logging if the deal had collapsed.

Bailey, on the other hand, emphasized the dangers to the Headwaters Grove without protection under the deal. In particular, she called attention to Maxxam's efforts to use the state-level salvage exemption to allow logging operations in the old-growth groves. EPIC had not been able to stop these operations entirely through litigation, but Maxxam had been limited primarily to removal of downed trees, while the old growth was excluded from the logging. Bailey said she believed that Maxxam would have begun salvage operations in the Headwaters Grove if the deal had collapsed.

Such operations would have certainly faced numerous hurdles including litigation, direct action, and the possibility of intervention by state or federal agencies. Even if they had been able to proceed, those operations would likely have been limited to downed trees. Still, the removal of downed trees would have been detrimental to the forest ecosystem and would have undermined the relatively pristine character of the grove. Maxxam probably would not have been able to cut the standing old-growth under the salvage provisions, but they undoubtedly would have sought to do so through other means. For example, Congressman Riggs had suggested the possibility of legislation to reduce the protections for murrelet habitat under the Endangered Species Act. Ultimately, without public acquisition, Headwaters Grove would have continued to be at risk from logging. The deal provided a greater degree of certainty that the grove would not be cut. But the various Headwaters groups weighed those risks and benefits differently, both in relation to the confidence they had in nonlegislative forms of protection and also on potential harms they saw as resulting from other aspects of the deal.

The cost of the acquisition was another source of concern. Between the federal and state portions, including the Owl Creek and Grizzly Creek additions by California, the total cost of the deal was $480 million. One issue for Headwaters activists was the question of what Hurwitz would do with the large influx of money. The steelworkers in particular were concerned that this funding would strengthen Maxxam in its lockout of workers at its Kaiser plants. Activists also worried that the funds could lead to Maxxam acquiring more forest lands for logging or engaging in other environmentally harmful activities elsewhere. Overall, the greatest frustration was that the price being paid for the land did not reflect the fact that Maxxam ostensibly could not log much of the area acquired due to the murrelet protection. In the deal negotiations, the land was valued as though Maxxam would be able to log it all. This dramatically increased the cost of acquisition.

The payment also set a bad precedent for endangered species protection. If the owner of endangered species habitat could get the government to purchase the land at a price that did not reflect the protections of the Endangered Species Act, that prospect was likely to encourage landowners to "ransom" imperiled habitat. Moreover, in the eyes of critics, the supporters of the deal were undercutting the legitimacy of the ESA for other environmental campaigns. Ken Miller stated, "My position was that the Headwaters deal was undermining the Endangered Species Act, and that the failure to oppose it on those grounds was a surrender to another attack on the Endangered Species Act."

A final but often overlooked component of the deal that also had implications for the Endangered Species Act was Maxxam's private property takings lawsuit against the federal government over the Owl Creek case. Federal officials claimed that the takings case was a key impetus for their efforts to negotiate a deal. However, none of the Headwaters activists interviewed saw the takings case as a legitimate danger. If anything, some of the critics of the deal saw it as regrettable that the case was dropped. Miller pointed out that "you couldn't ask for a better villain in court" than Hurwitz and Maxxam. The Maxxam lawsuit took place within a context in which the opponents of endangered species protection were using the takings issue as a primary basis for their attacks on the act. ESA opponents claimed that small property owners were harmed by the protection of obscure, uncharismatic species. If the Maxxam case had gone forward, it would have put a different face on the takings issue. On one side, there would be Maxxam, with a decade of negative publicity around the Headwaters issue, as well as links to junk bond scandals, corporate takeovers, the savings and loan crisis, and poor treatment of workers. This would be juxtaposed against the protection of ancient redwoods and imperiled birds, both of which enjoyed broad popular appeal. In other words, not only was the Maxxam takings case questionable in its legal merits, but the case was also likely to created bad publicity for ESA opponents.

It is not clear how far Maxxam would have actually been willing to pursue its takings claim. If Maxxam had continued with the suit, they could have wound up with a court ruling against them, which would have fortified the protections on the old growth in Headwaters and also created a strong legal precedent for ESA enforcement elsewhere. Of course, there was always the risk, however slim, that Maxxam could have won. By having Maxxam withdraw the suit, the federal government avoided this risk. But in doing so, they also gave undue legitimacy to Maxxam's claim by treating the halt of the suit as an important component of the deal negotiations. Josh Kaufman said, "Giving in to it was worse than testing it in court. And that's what happened. Now it is open season for all blackmailers."

Beyond the actual provisions of the deal, it is also important to consider its impact on forest protection organizing work in the region. While the deal had not protected most of 60,000 acres sought by the HFCC, its advocates contended that the campaign could still work to protect the remaining acreage after it was completed. However, this perspective downplayed the barriers to further victories created by the deal. As already noted, the habitat conservation plan and the mutual defense pact made litigation much more difficult. But perhaps more important, the deal had the effect of demobilizing the campaign. In the wake of the deal, the mainstream media interest in Headwaters declined, seeing the issue as more or less resolved. When activists tried to raise concerns over the ongoing threats to the forests, they were pilloried in the press. For example, one

San Francisco newspaper editorial began, "It was inevitable, we suppose, that when the ancient redwoods of Headwaters Forest were finally saved, a few boo birds would remain. Their tiresome chirping almost makes you want to become a greedy industrialist."[53]

The reduced media coverage made it more difficult to maintain public interest in the issue. This problem was further exacerbated by the fact that Headwaters Grove had served as the centerpiece of the campaign's public outreach up to this time. With the grove protected under federal ownership, public mobilization became more difficult. Funding also declined as some foundations shifted their grantmaking to other issues. As Bailey noted, "It is hard to raise money on a saved poster child." As a result, key grassroots groups went through a steep drop in funding and a subsequent downsizing. EPIC's total revenue dropped by almost half between 1998 and 1999. Immediately following the deal, Cecelia Lanman stepped down from her leadership position, and EPIC lost much of its staff in the years that followed. BACH managed to retain one part-time staff position, but its mission shifted from primarily mobilizing urban constituencies for protests to a focus on public input in the government's management of the newly created Headwaters Forest Reserve. And following the demise of the HFCC, Trees Foundation moved away from large-scale campaigns to focus on small-scale local conservation and restoration projects. Together these effects from the deal deflated the campaign. In Karen Pickett's assessment, "If the deal had been killed, the campaign wouldn't have had the legs knocked out from under it, which is what the deal did to the campaign. And in fact, it would have gotten stronger and bigger."

For all of these criticisms, it is worth asking whether there was any viable alternative to the deal for protecting the forest. Grassroots activists saw at least two other paths. Cecelia Lanman pointed out that the deal would have been unnecessary if the Clinton administration had chosen to vigorously enforce the Endangered Species Act:

> The crucial time is when we had this precedent in court, and the Department of the Interior decided to cave in to this takings and private property argument. To me, that is the crucial moment because that pretty much told us that we weren't going to get what we wanted. That's what started the deal making. I would have preferred to see the government saying, "Hey, you can't destroy endangered species habitat; I don't care whether you privately own these trees or not."

That would have been the ideal, of course. There wouldn't have been a deal in that case. It would have been the federal government enforcing the law in regards to what Hurwitz was doing there.

Ken Miller of Humboldt Watershed Council saw another path for protecting the forest that did not require a change of behavior by the Clinton administration. A key facet of Miller's alternative to the deal involved the financial pressures from Maxxam's ongoing and extensive debt from the Pacific Lumber takeover, combined with its reduced income due to environmentalists' successes in curtailing Maxxam's logging. Maxxam depended on the deal to provide a large infusion of cash to avoid defaulting on its debt, as well as the habitat conservation plan to provide assured levels of logging for making future debt payments. Miller lamented, "Hurwitz was very vulnerable. He had a plan and financially it was a house of cards. We allowed him to complete that plan." Without the deal, Miller believed,

> Hurwitz would have been in a very much weaker negotiating position with respect to his assets. He had creditors. He had high interest debt that he refinanced. He owed a lot of money—$600 million. And he was being stopped [from logging]. He couldn't get to his old growth. The murrelet was listed. The salmon was getting listed. So there were problems for him without an HCP. He needed an HCP. And his negotiating position in the HCP could have been very different had the Sierra Club just said, "Wait a minute, we're not going to go for this deal with this kind of HCP."

In Miller's account, if the Sierra Club had instead thwarted the deal, Hurwitz would have been in a much weaker position in negotiating for an independent habitat conservation plan; "Hurwitz would have been forced then to agree to an HCP with a lot less cutting, or he would have been fighting tooth and nail to get anything. He needed the HCP to get at his habitat." Furthermore, the reduced cutting under a stricter HCP would have created a financial crisis for Hurwitz that could have forced Maxxam to sell off the PL lands.

What Miller described was a hypothetical scenario. We will never know for certain whether it would have succeeded. Nonetheless, it is still significant because it illustrates that there were other options for protecting the forest outside of the framework of the deal. It thus serves as a useful

corrective to conventional forms of political analysis that evaluate campaign success primarily in terms of legislative outcomes. From such a perspective, Headwaters could only be saved through the actions of politicians
and the Headwaters deal was the last chance for a political solution to the
issue. If there was no alternative to the deal, then the best strategy for
environmentalists would be to try to improve it as much as possible. This
is the approach that Kathy Bailey and the Sierra Club ultimately championed. Rather than simply accept the deal, they used the California state
funding process to leverage notable improvements. Within that framework, they accomplished about as much as could have been hoped for.
And while Bailey acknowledged its shortcomings, those shortcomings
were outweighed by the belief that the deal was the only viable way to
save the forest.

In this approach, real power lay in the ability to influence the deal negotiations. From this perspective, there were those who did the substantive
work of political negotiation and there were outsiders who could criticize
the outcomes of those negotiations but who did not actually have much
influence on those outcomes. However, this dichotomy overlooks that the
critics of the deal were more than just naysayers. Grassroots activists such
as Ken Miller offered plausible alternative routes to greater forest protection outside of the framework of the deal. While Bailey's approach represented the best that could be accomplished within that framework, at the
same time this approach precluded alternatives outside of that framework
that potentially could have been more effective. Thus the Headwaters deal
had a complex character as simultaneously a partial victory, a partial setback, and a missed opportunity.

After the Deal

The finalization of the deal was a decisive moment for the Headwaters
campaign, and afterward, the campaign unraveled. That said, grassroots
activism to protect the forests on Pacific Lumber's property did not end
with the deal. While many of the tactics that had been particularly effective earlier in the campaign were less successful after the deal, new tactics
emerged in response to the changing conditions. Activists pursued a variety of tactics simultaneously. With so many facets, it is difficult to describe
the events of the postdeal period chronologically. Instead, the following
account of that phase of the campaign is organized in terms the tactics

used—direct action, alliance with labor, litigation, political approaches, and alternative approaches.

Direct Action

The most successful direct action of the Headwaters campaign in the wake of the deal was Julia "Butterfly" Hill's tree-sit. It began more than a year before the deal was finalized. At that point, Julia Butterfly had only recently arrived in Humboldt County to help protect redwoods. When she arrived, she had no established connections with any Headwaters groups and no experience with direct action activism. However, Earth First! provided a base camp where she could meet people who would incorporate her into a tree-sit and teach her the technical climbing skills it required. The tree-sit in the old-growth redwood that Julia Butterfly named Luna began like other tree-sits—two or three volunteers spent a few days in the tree until replacement volunteers could be found, in a repeating cycle. However, as winter approached, it seemed unlikely that they would be able maintain a supply of new tree-sitters for Luna. In December 1997, Butterfly decided to embark on a longer solo tree-sit to fill this gap in coverage. She originally intended to stay in Luna for about a month, but wound up remaining in her tree for just over two years. During that time she faced severe weather during the rainiest winter in more than a century, as well as harassment from Maxxam—security guards placed at the base of the tree to try to prevent her support team from bringing her food; spotlights and noise-makers to keep her awake at night; a helicopter that almost blew her out of the tree; and logging operations precariously close to Luna. While most of the other trees in the area were cut, Luna remained standing and Butterfly remained in Luna.

She continued her tree-sit even after the deal was finalized. Her tree was outside of the new Headwaters Forest Reserve and not protected by the provisions of the deal, so she stayed in the tree and spoke out publicly against the deal. Her tree-sit was the best hope for giving a "face" to the parts of the forest that had not been protected. Luna became as well known as Headwaters Grove, and there was a similar sense that public would simply not allow Maxxam to log Luna because of the attention raised by Butterfly's tree-sit. The duration of her action was unprecedented. Prior to that, the longest tree-sit had been ninety days.[54] While Butterfly's vigil was initially ignored by the media, it gradually began to attract more interest and eventually garnered national and international coverage. She wound up receiving more media attention than other any figure in the Headwaters campaign and became one of the most well-known environmental activists in the United States.

From a cell phone in her perch, Butterfly talked directly with reporters and even participated in a debate with Pacific Lumber's president on CNN. She also got media coverage from sources that otherwise would not pay attention to the Headwaters issue, such as *Good Housekeeping* magazine, which nominated her as one of the most admired women in America. All of the publicity put increasing pressure on Maxxam to negotiate with her. Finally, soon after the second anniversary of Butterfly's assent into Luna, Maxxam signed a conservation easement to preserve Luna as well as a two-hundred-foot buffer area in exchange for Butterfly agreeing to end her tree-sit and not engage in further tree-sits on Maxxam property. On December 18, 1999, Butterfly climbed down from Luna and her feet touched the ground for the first time in 738 days.

The protection of Luna was one of the few victories for direct action in the postdeal era.[55] Butterfly's tree-sit inspired a proliferation of long-term tree-sits on Maxxam lands, some lasting a year or more. However, they did not generate anywhere near as much media attention; the tactic had lost its novelty and newsworthiness. At the same time, Maxxam became more effective in extracting protestors from the trees. While there had been twenty-two tree-sits on Maxxam lands in spring 2003, there were only three remaining by fall 2004.[56]

Alliance with Labor

The coalition between the United Steelworkers of America and Earth First! continued to grow rapidly in the period immediately following the deal. In spring 1999, Earth First!ers invited the steelworkers to join them at an environmental conference in Eugene, Oregon. There they met longtime environmental leader David Brower, and together they hatched the idea for a broader labor-environmental coalition to be known as the Alliance for Sustainable Jobs and the Environment.

The new alliance participated in protests at the Maxxam shareholders meeting in Houston in 1999. Although Headwaters activists had been attending the meeting for years, the presence of 150 "burly" steelworkers gave the event a decided differently flavor.[57] The striking Kaiser workers directly confronted Hurwitz inside the meeting. They were joined by the mother of David "Gypsy" Chain and representatives of

Rose Foundation and As You Sow Foundation, who had prepared share-holder resolutions to reform Maxxam's practices. Although the resolutions did not pass, the event was one of the most diverse and dynamic protests of the postdeal era.

Out of that meeting, the environmentalists and union activists developed a joint declaration on the social and ecological harm from corporate malfeasance known as the "Houston Principles," which would become a key document of the alliance. They then published a full-page ad to announce their alliance: the headline declared, "Have You Heard the One about the Steelworker and the Environmentalist? Lots of Companies Wish We Were Joking. We're Serious."

That fall, the alliance prepared for joint participation in the protests of the Seattle World Trade Organization (WTO) meeting in November. That event would turn out to be one of the most significant demonstrations in the United States in recent history, drawing together tens of thousands of union members, environmentalists, and social justice activists. Mass direct action in the streets of Seattle shut down the meeting of the WTO for a day, and the WTO ultimately adjourned without reaching any new agreements and with the future of the organization in question.

Subsequent accounts of the protests would highlight the importance of the cooperation between the labor unions and environmental organizations in protesting the WTO, often referred to in terms of "Teamsters and turtles." While that phrase provided a memorable alliteration, it obscured the central role played by the Earth First! forest activists, the steelworkers union, and the connection between them that had already developed through the Headwaters campaign. The activist blockades of the streets around the WTO meeting used the lockbox techniques that had been refined by Earth First!ers in the Headwaters campaign. The leaders of the main AFL-CIO unions were reluctant to associate themselves with these actions, and initially the union members were kept away from the street protests around the WTO meeting center. However, as the direct action activists faced intense repression from the Seattle police in the form of tear gas, pepper spray, rubber bullets, and mass arrests, the steelworkers proved to be a crucial bridge between those activists and the more moderate unions. Particularly important were the bonds that rank-and-file steelworkers such as John Goodman and Don Kegley had formed with the environmentalists while working on the Maxxam campaign.[58] Having worked closely with Earth First!ers for more than a year, the steelworkers were unwilling to stand by idly while their friends were hurt or arrested by the police. As a result, unions wound up playing a more active role in supporting the street protests, which further intensified the pressure on the WTO delegates and contributed to the collapse of their meeting.

Following Seattle, the labor-environmental alliance continued to develop and the Alliance for Sustainable Jobs and the Environment became a formal organization. However, in subsequent years, the growth of that entity would be constrained by a dearth of foundation funding. Meanwhile, the Kaiser workers' strike went almost two years until the steelworkers were able to secure a favorable contract from Maxxam in fall 2000.[59] Karen Pickett read Don Kegley's account of the steelworkers' celebratory return to the Kaiser plants to environmental leader David Brower just days before his death at age eighty-eight. The emerging alliance between environmentalists and labor had filled much of Brower's final year and had re-inspired his activism. At the end of a lifetime of environmental advocacy, Brower said that alliance "made it possible for me to dream again."[60]

Litigation

The conditions attached to the Headwaters deal—particularly the habitat conservation plan and the mutual defense pact between Maxxam and the government—markedly reduced the success of environmental lawsuits in the postdeal era. As a result, the overall effectiveness of the Headwaters campaign was notably diminished. Nonetheless, Headwaters activists continued to explore new ways to use litigation to protect the forest.

Following the finalization of the deal, the Headwaters groups immediately prepared to litigate against the federal habitat conservation plan and the state-level sustained yield plan. Under California law, the lawsuit against the sustained yield plan had to be filed within thirty days, so that became the first priority. By the end of March 1999, EPIC and the Sierra Club brought suit against the plan, while steelworkers filed companion litigation over the long-term economic implications of that plan. However, the plaintiffs did not get a ruling until 2003. The first judge assigned to the case withdrew after almost a year, citing a relationship with a Maxxam executive. The judge who replaced him stayed on the case for another eighteen months before disclosing that he too had a relationship with a Maxxam employee and removing himself from the case.[61] The lawsuit finally went to trial under a third judge four years after the case was filed. During this time, Maxxam was able to proceed with its logging under the plan unhindered. Ultimately, the plaintiffs won the case. However, while the judge agreed with the plaintiffs' claims, when it came time to determine what actions should result from his ruling, he chose not to stop any of Maxxam's logging.

The sustained yield plan litigation was originally seen by environmentalists primarily as a warm up to the more significant legal challenge to the federal habitat conservation plan. However, the HCP litigation got bogged down before any case could be filed. For more than four years, EPIC's limited resources were swamped by the sustained yield case. A litigator from Earthjustice initially took the HCP case, but withdrew from it a year later, and funding could not be found to hire other lawyers for what would be a difficult and expensive case (especially given the mutual defense pact). So the HCP went unchallenged until EPIC filed suit against it in late 2004.

Through those previous five years, Maxxam's HCP remained in effect and EPIC was prevented from using the Endangered Species Act to stop Maxxam's continued logging of endangered species habitat not protected under the deal. In 2005, Cynthia Elkins of EPIC noted, "There hasn't been a successful challenge against a timber harvest plan since Pacific Lumber has had its HCP." Instead, Elkins turned to innovative new forms of environmental litigation using the federal Clean Water Act to challenge the water pollution produced by siltation resulting from logging.

Following the completion of the Headwaters deal, Maxxam also faced two high-profile lawsuits involving attorney Steve Schectman. The first was a suit on behalf of residents of the town of Stafford whose homes were damaged or destroyed by the 1996 New Year's Eve landslide that began in a Maxxam clearcut on a steep slope. Schectman, a successful litigator from San Francisco, had moved his family to Humboldt County in 1998 to work on that case. There he also became an attorney in the wrongful-death suit brought by the parents of David "Gypsy" Chain after authorities refused to prosecute the logger responsible for his death. With these and other lawsuits against Maxxam on his plate, Schectman offered to take on several Humboldt activists as legal apprentices for him on the Maxxam cases. After four years of apprenticeship, they could take the bar exam and become lawyers without attending law school. The plan was to produce a new generation of activist lawyers while simultaneously providing Schectman with low-cost, motivated assistants for his cases.[62]

However, Schectman's grassroots firm became overwhelmed by the workload of addressing both the Stafford and Gypsy suits simultaneously, and Schectman went deeply into debt preparing these cases. With Schectman's resources strained to their limits and Maxxam having a well-funded legal team, both cases wound up settling before trial. In March 2001, just three days before the Stafford trial was scheduled to begin, Maxxam agreed to pay the residents $3.3 million. Then in October, a settlement was reached in Gypsy's wrongful-death case for $825,000. While both cases were ostensibly victories for the plaintiffs, the settlement amounts were significantly less than expected and Maxxam was able to avoid the potential embarrassment from high-profile trials.[63]

Two of the most notable courtroom successes of the postdeal era involved civil rights cases: the pepper spray suit and Judi Bari and Darryl Cherney's bombing case. The activists' legal teams in both cases were headed by activist-attorneys Tony Serra and Dennis Cunningham, and both cases took a long time to reach successful conclusions. The pepper spray case—in which eight forest activists sued Humboldt County after the sheriff's deputies swabbed their eyes with pepper spray while they were locked-down in Headwaters direct actions in 1997—went through three trials over seven years before reaching a final verdict. While the activists' lawyers had asked the jury to award their clients thousands of dollars in damages for their pain and suffering, the activists had made clear throughout the trial that they were not pursuing the case for the money. The jury responded with an award of one dollar for each of the plaintiffs. Despite the small damage award, the suit had effectively stopped the Humboldt sheriff from further misuse of the pepper spray against forest activists. The trials also generated extensive media coverage and newspaper editorials that were generally sympathetic to the Earth First!ers right to commit nonviolent civil disobedience in defense of the forests.

Even longer and more successful was Bari and Cherney's lawsuit stemming from the pipe bomb that was placed under Judi Bari's car seat in 1990. They sued the FBI and the Oakland police for civil rights violations stemming from the authorities' portrayal of them as terrorists responsible for making the bomb despite evidence to the contrary. The lawsuit was filed in 1991. The litigation gave the activists the opportunity to depose the agents involved in the case starting in 1994. Through these depositions, it was discovered that the FBI's bomb expert on the case had organized a bomb investigator's training course (dubbed "bomb school") in Humboldt four weeks before the bomb exploded under Bari. As part of that bomb school, the FBI detonated pipe bombs inside the passenger compartments of several cars for participants to practice their responses. This practice bombing took place on forest land owned by a timber corporation and it was attended by at least four of the FBI agents who were involved in responding to the pipe bomb in Bari's car just weeks later.[64] The lawsuit investigation also produced Oakland police photos showing that the gapping hole from the explosion in the floor of Bari's car was clearly centered under her seat, not in the back of the car as the FBI agents had reported. And during the depositions an FBI bomb expert from Washington, DC testified that the pipe bomb had been triggered by a very sophisticated motion-sensitive device, which contradicted the image that the FBI and police had presented to the media of the activists accidentally bombing themselves.[65]

However, it would be a long wait before any of these findings could be presented in a courtroom. The defense attorneys filed motions that stalled the case for many years. In fall 1996 and spring 1997, knowing that she was facing terminal cancer, Bari spent the last months of her life preparing the materials for the case so that it could continue after her death. Just weeks before she died, she videotaped her deposition so that it would be available if the case ever came to trial. She also made Darryl Cherney promise that he would not settle the case. Over the next five years, Cherney and the legal team continued to pursue the case on scant resources.

Finally, the case went to trial in April 2002. Coming soon after the World Trade Center attacks, the trial took place at a time when concerns over terrorism were particularly charged and where the Patriot Act had recently reduced oversight of the FBI. Nonetheless, after a five-week trial and eighteen days of deliberations, the jury ruled in favor of Bari and Cherney and awarded them a total of $4.4 million. Then in April 2004, the FBI and the city of Oakland agreed to not pursue appeals and to pay $4 million. It was the largest amount ever paid by the FBI for a violation of civil rights.

While this case, like the pepper spray suit, did not directly protect any trees, both lawsuits created greater legal protections for the rights of the Earth First! activists engaged in defending forests. Moreover, both cases provided rare and energizing victories for Headwaters activists during the postdeal era in which the overall effectiveness of environmental litigation had been markedly diminished.

Political Approaches

The election of President George W. Bush in 2000 combined with the dominance of the more antienvironmental wing of the Republican Party in Congress largely precluded any possibility for additional federal action to protect Headwaters after the deal. Meanwhile, in California, Democrat Gray Davis had become governor just as the deal had been completed. Initially, the arrival of Davis was a source of hope for some environmentalists because he had courted the environmental vote during the election and made a campaign pledge to ensure that "all old-growth trees are spared from the lumberjack's ax."[66] Davis appointed some environmentalists to positions in his administration. For example, Louis Blumberg, a staff member of The Wilderness Society, became the spokesman for the California Department of Forestry. However, this appointment proved to be of questionable benefit to forest activists, since Blumberg's job required him to defend the Davis administration's policies from environmentalists' criticisms as that administration grew increasingly close to the timber industry.

An early indication of what was to come was Davis's appointment of former Maxxam executive Barry Munitz to head his transition team immediately after the election. Soon after taking office in 1999, the Davis administration proposed increasing regulation of the timber industry, but on the same day that the planned reforms were announced, the largest timber company in California hosted a fundraiser for Davis that was attended by many timber industry executives. Soon thereafter, the Davis administration backed away from implementing its logging regulations.[67]

Then in January 2000, state wildlife and forestry agency staff were called to a meeting purportedly about the implementation of the Headwaters habitat conservation plan. There they were reprimanded by the governor's deputy chief of staff, Susan Kennedy, for not approving logging at a rate that satisfied Maxxam/PL. The meeting was also attended by Pacific Lumber's executive vice president Jared Carter. In front of Carter, Kennedy instructed the staff to "bend over backward" to expedite the approval of Maxxam timber harvest plans. One staff member at that meeting described it as a "public whipping," adding that "the company came to the administration and complained; so the governor's office beat up on us a bit."[68]

Despite these developments, Headwaters activists hoped to hold Davis to his pledge to end the logging of old-growth trees in California. To that end, they formed the Campaign for Old Growth, with EPIC activist Sue Moloney serving as executive director. When the governor was not responsive, they turned to the state ballot initiative process. In June and July 2002, they mobilized an all-volunteer effort to gather signatures to get an initiative on the November ballot that would end the logging of old growth. However without the financial resources of the 1990 Forests Forever campaign, they were not able to gather enough signatures in time. But then, State Senator Don Perata stepped in and introduced the language of the initiative as state legislation. Meanwhile, that fall, Sue Moloney embarked on a hunger strike to pressure Governor Davis to act on his old-growth pledge. Moloney sat at the capitol on a juice-only fast for fifty-two days, but the governor was unswayed. In summer 2003, Perata's old-growth protection bill was passed by the state senate, but it was then delayed in committee in the assembly. The Campaign for Old Growth rallied celebrities such as Pierce Brosnan and Martin Sheen to the issue, and by summer 2004, the activists believed that they were close to obtaining the support of a majority in the assembly. However, the speaker of the assembly did not let the bill come to a floor vote, and so it died at the end of the legislative session.

Alternative Approaches

One of the great strengths of the Headwaters campaign was its use of innovative approaches to put pressure on Maxxam. Headwaters activists did not rely solely on familiar environmental agencies such as the California Department of Forestry and the U.S. Fish and Wildlife Service. Such agencies were important sites for struggle for the Headwaters campaign, but timber corporations such as Maxxam were experienced with applying counterpressure on them. Therefore, Headwaters activists did not limit themselves to those venues and looked for new and unexpected ways to challenge Maxxam.

The best known of these alternate avenues was the debt-for-nature tactic, where potential fines against Maxxam by the Federal Deposit Insurance Corporation (FDIC) and the Office of Thrift Supervision (OTS) might be converted into additional acquisitions of Maxxam forest land. Following the completion of the Headwaters deal, the Rose Foundation continued to push the debt-for-nature approach as a means to protect more of Headwaters Forest, hoping that 10,000 to 20,000 acres of redwoods could be added to the reserve this way. Then in 2000,

Maxxam countersued the government, alleging that the FDIC had improperly financed the OTS's suit against Maxxam in order to bolster its own case. In 2002, the OTS settled with Maxxam for $206,000, a tiny fraction of the $820 million it had originally sought. Under that settlement, Maxxam made no admission of wrongdoing. Soon after, the FDIC dropped its case against Maxxam. However, Maxxam continued its countersuit, and in 2005, a federal judge ordered the FDIC to pay Maxxam up to $72 million for the latter's legal fees in fighting both the FDIC and OTS cases.[69]

Less than a month after OTS settled its case in 2002, Maxxam faced a new pressure from an unexpected corner—the Humboldt County district attorney. In the November election, Paul Gallegos, a former defense attorney, unseated the incumbent district attorney who had held that position for twenty years. Gallegos did not run on an environmental platform, but he filed a civil fraud suit against Maxxam soon after taking office. Based on information provided by Ken Miller of the Humboldt Watershed Council, Gallegos charged that Maxxam had supplied fraudulent information to the California Department of Forestry (CDF) in the process of preparing its sustained yield plan as part of the Headwaters deal. The main area of concern was a watershed known as Jordan Creek. Initially, Maxxam had reported to the CDF that only 15 percent of the landslides along the creek were the result of logging. But two days before the end of the environmental review period, Maxxam changed those figures to show that logging was responsible for 60 percent of the landslides. However, Maxxam only reported the corrections to a local CDF official rather than the CDF headquarters, and thus the new figures were not included in the sustained yield plan analysis. As part of his declaration for Gallegos's lawsuit, the director of CDF at that time stated that he had not received the corrected figures. Furthermore, he indicated that if he had received those figures he would have sent the plan back for further environmental review.[70] Gallegos contended that Maxxam had submitted false data on landslide risks in order to be allowed to log 100,000 trees on unstable slopes. At the time the suit was filed, about 30,000 of those trees had been cut. Gallegos was seeking $2,500 per tree in civil penalties for a total potential cost to Maxxam of at least $75 million.[71]

Then, less than six weeks into his first term, a recall campaign was launched against Gallegos. That campaign would prove to be the most expensive in Humboldt County history, with Maxxam contributing 80 percent of the funds for the recall.[72] Headwaters activists worked on the campaign to fight the recall, and in March 2004, voters rejected the recall by a margin of 61 to 39 percent—a larger percentage than had voted for Gallegos in the previous election. The recall election outcome was a victory for Gallegos and allowed him to continue work on the Maxxam lawsuit. However, in June 2004, the judge dismissed the case, agreeing with Maxxam's contention that, under the First Amendment, a corporation does not need to tell the truth when lobbying the government.[73]

While neither of the previous two examples of alternate approaches directly protected any trees, they both forced Maxxam to defend itself from unexpected angles and generated extensive negative publicity about Maxxam. An alternate approach that did directly protect trees was the Humboldt Watershed Council's campaign to reduce Maxxam's logging levels through the intervention of California's water board. The board had not been a signatory to the Headwaters deal. As such, it was not obligated to support the logging levels granted by the deal and was not involved in the mutual defense pact. Moreover, the water board had the power to regulate activities that harmed water quality, including the siltation and debris in waterways resulting from Maxxam's logging. Accordingly, it represented a key leverage point for activists in the postdeal era.

The Humboldt Watershed Council led the efforts to move the board toward what Ken Miller described as a "science-based policy of regulation for recovery." Pressure around water quality issues led to a moratorium on new logging in the Freshwater Creek (1999–2001) and Elk River (1999–2002) watersheds. Ultimately, Maxxam's logging in these areas would depend on getting a new type of permit that was being developed by the regional water board, called Watershed-wide Water Discharge Requirements. Under pressure from Maxxam, the executive officer for the regional water board granted interim permits to the company until the discharge requirements could be finalized. These interim permits allowed Maxxam about 50 percent of the logging in the Freshwater and Elk River watersheds that CDF had permitted. Maxxam declared that if it did not receive permission to cut 100 percent of its timber harvest plans, then Pacific Lumber would not be able to make the debt payments to its creditors and might have to file for bankruptcy. In March 2005, the regional board raised the amount of logging it would allow Maxxam to 75 percent of the timber harvest plans. Then in April, Humboldt Watershed Council successfully appealed to the state-level Water Resources

Control Board, which returned the logging levels in those watersheds to 50 percent until the permit was finalized.[74]

Maxxam continued to claim that Pacific Lumber would face bankruptcy if it was not allowed to log all of its timber harvest plans. However, a water board researcher concluded that the company's debt problems were entirely of its own making. His study found that Maxxam had "sucked more than $724 million from its subsidiary by cutting trees at unsustainable rates."[75] Yet at the same time, the debt load for Pacific Lumber remained about the same as it had been right after Maxxam's buyout of the company. Moreover, because of the high debt load Hurwitz arranged for Maxxam's timber subsidiary under a 1998 restructuring, the water board researcher calculated that it might not be possible for PL to pay off that debt even if it cut every one of its trees.[76]

Pacific Lumber ultimately did file for bankruptcy in January 2007. The bankruptcy process was completed in summer 2008 with the Mendocino Redwood Company purchasing PL's lands and mill. The Mendocino Redwood Company eschewed clearcutting and promoted more environmentally responsible logging practices, so its bid to acquire Pacific Lumber was supported by EPIC. Charles Hurwitz and the Maxxam Corporation would no longer shape the fate of the remaining redwoods.

While the Headwaters campaign in the decade following the deal struggled against difficult hurdles and did not produce nearly as many dramatic achievements as it did prior to the deal, the grassroots activists persisted. And in the end, more than twenty years after Headwaters activists had first called for "Maxxam out of Humboldt County," their goal was finally realized.

The Legacy of the Headwaters Campaign

The Headwaters campaign was unusual among forest protection campaigns in that it was led primarily by small grassroots groups. These groups did not have the sizeable resources and ready access to politicians that characterize the national environmental organizations. Nonetheless, the campaign was able to turn the fate of Headwaters into a national issue and to compel the federal government to intervene. Innovative litigation forced the issue by halting much of the old-growth logging. At the same time, thousands of people participated in demonstrations in support of Headwaters, leading to the largest mass civil disobedience for forest protection in U.S. history. This mass mobilization contributed to the willingness of the federal and state governments to spend almost a half billion dollars to try to resolve the issue. In the end, Headwaters Grove and three smaller groves were acquired for protection under public ownership, and the other main old-growth redwood groves within Headwaters Forest were secured from logging for fifty years through the Headwaters deal. Even after the deal, activists were able to secure some additional protections for trees on Maxxam's land, particularly through California's water board.

The benefits of the Headwaters campaign also extended beyond the redwood forests. EPIC's courtroom victories created important legal precedents for environmental protection at the state and federal levels. Perhaps most significant, the marbled murrelet case established a firm basis for federal protection of endangered species habitat on private lands. Likewise, the successful civil rights cases involving Judi Bari and Darryl Cherney and the pepper spray victims helped to limit police and FBI repression against environmental activists. The Headwaters activists were also quite innovative in exploring nontraditional means of putting pressure on corporations. While not all of these approaches were directly successful, the experiments provided useful lessons that could then inform the tactics of other campaigns. For example, Rainforest Action Network's redwood boycott served as an important first step in the development of a broader set of market-based campaigns that would produce significant international forest protection victories in the 2000s.

The Headwaters campaign was also at the heart of a fundamental transformation within Earth First!. Judi Bari and Darryl Cherney were central figures in this shift through their efforts to build alliances and to explore more inclusive forms of direct action, such as Redwood Summer and the September 15th protests. As a result, Earth First!'s image changed from being primarily identified with covert monkeywrenching in the 1980s to more broad-based civil disobedience in the 1990s. The mass trespasses, road blockades, lock-downs, and tree-sits of the Headwaters campaign brought many new participants to direct action. And the campaign provided an important training ground for the next generation of environmental activists, including Julia Butterfly. Furthermore, the connections built between Earth First!ers and steelworkers played a crucial role in the success of the protests at the Seattle WTO and the growth of labor-environmental alliances.

Innovations and Constraints in the Headwaters Campaign

How was the grassroots-based Headwaters campaign able to accomplish all of this? To address this question, I examine the role of movement culture, tactics, strategy, organization, funding, and political conditions in shaping the campaign. I then conclude by assessing the relationship of the grassroots groups to the national environmental organizations and its implications for the outcome of the campaign.

One distinctive feature of the Headwaters campaign was the large number of grassroots groups involved in that campaign. This abundance was nurtured by the extensive counterculture community in northern California. The values of this community encouraged local support for grassroots environmental groups such as EPIC, which sustained these groups even before grantmaking foundations were playing a significant role.[77] (This context helps to explain why EPIC formed earlier than many of the other grassroots biodiversity groups that will be profiled in subsequent chapters. This enabled EPIC to serve as a role model for groups such as the Center for Biological Diversity.) The northern California activists were further emboldened by their contact with Earth First!'s radical movement culture, which fostered a more confrontational approach to forest activism. The Earth First! community also provided an accessible entry point and training ground for young activists who would subsequently play key roles in other Headwaters groups. The combination of northern California's rural counterculture and Earth First!'s movement culture thus provided fertile terrain for the growth of bold, new grassroots groups.

The new groups developed a remarkable array of tactics. Together they put continuous pressure on Maxxam from a variety of unexpected angles. Among the myriad tactics, innovative litigation initiated by EPIC proved to be particularly effective in halting logging and thus providing a strong impetus for Maxxam to enter into negotiations about preserving Headwaters.

For all of the tactical diversity within the Headwaters campaign, most of the groups shared an outsider strategy for protecting the forest. This is not to say that grassroots Headwaters activists never engaged in political outreach. At times, members of EPIC, Earth First!, and other grassroots groups traveled to Washington, DC and Sacramento to speak to government officials, but they did not have special access to these officials, nor did they see such access as the basis for their influence. Instead, they saw their power as rooted in their ability to directly block logging through litigation and direct action. As a result, their goals and their tactics were not constrained by the need to maintain access to politicians. As such, they were more likely to advocate for levels of protection based on the ecological needs of the redwood forest ecosystem, rather than limiting themselves to smaller demands that might seem more politically expedient in the short term but would not preserve the forest and its wildlife over the long term. Meanwhile, most of the national environmental organizations that embodied the insider strategy played little or no role in the Headwaters campaign, and when they did intervene, the results were problematic, as I will discuss later.

The grassroots Headwaters groups organized an informal network for communication and coordination that allowed the small groups to magnify their collective influence, providing an alternative to having a national environmental organization lead the campaign. The network of Headwaters groups using different tactics to put pressure on Maxxam from numerous angles resembled the "swarming" form of organization later used effectively by the WTO protesters in Seattle.[78] The success of the swarming approach depended on a decentralized network of groups that were autonomous but worked together in a coordinated manner. Such networks lacked a central leadership that could be readily co-opted or otherwise neutralized by opponents. However, funding pressures led to a centralization of the Headwaters network through the Headwaters Forest Coordinating Committee. Grantmakers wanted a central point through which to channel their Headwaters funding, and this in turn required a more formal coalition structure to distribute the funds and make collective decisions. The resulting organizational apparatus took up a lot of the organizers' time and sparked conflicts between Headwaters groups that had been working well together until that time. While some level of decentralized organization was useful, creating a more centralized formal coalition was ultimately detrimental to the campaign.

The role of grantmaking foundations in creating the more centralized version of the HFCC was illustrative of the Headwaters groups' complicated relationship to outside funding. This was also evident in terms of both the benefits and problems associated with Hal Arbit's sponsorship of the Forests Forever initiative. It is not surprising then that many Headwaters activists expressed ambivalence toward the role of funding in the campaign. From their experiences, funding was not simply a resource that increased the capacity of a social movement. It could also play a disruptive role, especially when funding levels changed quickly. Moreover, funding was often tied to increased influence from the funders, which, while well-intentioned, could inadvertently undermine the grassroots approach pursued by the Headwaters groups.

The campaign's relationship to Democratic politicians was also complicated. While the Democratic Party is usually seen as being an ally to environmentalists, Headwaters activists who were critical of the Headwaters deal were also very critical of the role played by the Clinton administration and Democratic senator Dianne Feinstein. They felt let down by the Clinton administration's failure to strongly enforce the Endangered Species Act in the face of Hurwitz's threats to the old-growth groves, and ultimately they felt betrayed by Clinton and Feinstein in regards to the final content of the Headwaters deal. Likewise, the interventions by Governor Davis's administration on behalf of Maxxam during the implementation of the deal were a further setback for Headwaters activists. This is not to say that Headwaters activists never benefited from having Democratic politicians in office. For example, Congressman Dan Hamburg and State Senator Tom Hayden were important allies, but they came to their elected roles already with close ties to grassroots activists. Thus, they were atypical, and they were on the margins of the Democratic Party. (This was perhaps best exemplified by Hamburg, who switched over to the Green Party after feeling abandoned by Democrats during his contested reelection campaign.) By comparison, the moderate Democrats who were ascendant to leadership roles in the 1990s, such as Clinton, Feinstein, and Davis, were, at best, unreliable allies whose actions (and inactions) wound up undermining the Headwaters campaign at key junctures.[79]

Ultimately, the most problematic relationship for the grassroots Headwaters groups involved a national environmental organization—the Sierra Club. For much of the campaign, most of the national environmental organizations that usually dominated forest protection efforts played little or no role in protecting Headwaters. In this regard, the approach of grassroots groups was to say "never mind the nationals" and find ways to be influential without those larger organizations. However, this approach ran into trouble when Club staff chose to intervene in the campaign at key junctures, including the Forest Forever initiative and, most notably, during the Headwaters deal. Executive Director Carl Pope personally intervened with the California Senate to get the state portion of the deal funding passed even though most of the grassroots Headwaters organizations were actively opposing the deal. If Pope had instead asked the senate president to block the bill, the deal most likely would have failed and the outcome of the Headwaters campaign might have been very different. At no other point would environmental groups have such singular control over the outcome of the campaign. And thus it offers a particularly well-defined moment to examine the relationship between a national environmental organization using an insider strategy and grassroots groups that relied primarily on an outsider strategy.

The Relationship between the Grassroots and the Nationals in the Headwaters Campaign

There are at least three interpretations of the intervention of the Sierra Club in the culmination of the Headwaters deal: first, the view that the Sierra Club staff had to rescue Headwaters, via the deal, from the limitations of the grassroots groups; second, the view that the Club and the grassroots groups played complementary roles that maximized the protection of Headwaters; and third, the view that the intervention by Club staff undercut the grassroots strategy and ultimately lead to less protection for the forests.

From the first perspective, the grassroots groups had become trapped by their own radical rhetoric in opposing the shortcomings of the deal. While all the groups agreed that the deal had flaws, to supporters, it offered the only available avenue for providing more secure protection for much of the remaining old-growth redwoods. The Sierra Club executive director therefore chose to intervene unilaterally to rescue the deal from the grassroots efforts to stop it. Ultimately, it was a paternalistic action that disregarded the judgment of many grassroots activists who had long been involved in the campaign and saw viable alternates to the deal.

A more complementary view suggests that the combination of grassroots groups critiquing the deal from the outside and the Sierra Club working on the inside of the political arena maximized the protections provided to Headwaters. From this perspective, the threat of the grassroots groups blocking the deal prevented the Sierra Club from compromising more and gave Kathy Bailey greater leverage in her negotiations to improve the deal. Although Karen Pickett was a strong critic of the deal, she also noted the way that the grassroots groups had strengthened the final outcome: "The things that were gained to make it incrementally better—stream buffers, that kind of thing that happened at the state level—wouldn't have happened were it not for the grassroots groups holding the hard line." The outsiders' efforts thus assisted the insiders in getting a better deal even though the outsiders opposed the deal. While this might be described as a complementary relationship, it was certainly an inadvertent one. Moreover, it may have been unnecessary.

The complimentary perspective, like the first perspective, depends on the assumption that the deal was the only way to protect Headwaters. However, that assumption may have been incorrect. Grassroots activists such as Ken Miller offered plausible alternative routes to even greater forest protection if the deal had been stopped. For them, the Headwaters deal was an unnecessary compromise that sacrificed protection for much of the forest outside of the old-growth groves, demobilized the campaign, strengthened Maxxam, and weakened the Endangered Species Act. From that perspective, the intervention of the Sierra Club in the passage of the state funding was not complementary. Instead, it undercut the strategy being pursued by most of the Headwaters groups. Speaking prior to the Pacific Lumber bankruptcy, Ken Miller exclaimed,

> [The deal] was a devil's bargain. And it undermined our power. It totally undermined it. People who say we couldn't have got [protection for Headwaters] otherwise—that's a self-fulfilling prophesy. . . . They took away our power by doing this deal. If we had been united as an environmental movement and as a community, this deal wouldn't have happened and Hurwitz wouldn't be here. We could restore fisheries and we wouldn't have lost our watersheds. We could have had a [sustainable] economy here.

There have been many examples of environmental campaigns whose outcomes were brokered by the national organizations. At the same time, there has been a growing critique of the shortcomings of the insider strategy used by the nationals. The Headwaters campaign offered a rare chance to see what could be achieved by groups using a different strategy. Up until the deal, the Headwaters campaign showed that grassroots groups could accomplish quite a lot without the involvement of the national environmental organizations. Then Sierra Club staff intervened to save the Headwaters deal over the objections of the grassroots. Thus, the outcome of the deal was a missed opportunity to see what a grassroots approach could achieve on its own if given the chance. Nonetheless, the grassroots activists persisted, despite the hurdles created by the deal, and their actions ultimately contributed to the end of Maxxam's reign in the redwoods.

4

TRANSFORMING A NATIONAL:
THE JOHN MUIR SIERRANS AND
THE ZERO-CUT CAMPAIGN

The Headwaters Forest campaign described in the chapter 3 differed from most other forest protection campaigns in two key ways. First, Headwaters Forest was on private lands, whereas most other forest protection campaigns focused on national forests and other public lands. While private timber companies still did the logging on national forests, the responsibility for the management of those lands and the development of timber sales was delegated to a federal agency, the U.S. Forest Service. However, the behavior of the Forest Service suggested that the agency was primarily concerned with "getting the cut out," even when doing so was harmful to the forest ecosystems.[1] In the 1970s, new laws such as the National Forest Management Act and the National Environmental Policy Act were created to provide greater oversight of the Forest Service, but initially the implementation of those laws was limited, and in the 1980s logging levels on national forests rose to nearly record levels. (For a history of these laws, see the appendix. For a graph of national forest logging levels, see figure 4.1.) Nonetheless, these laws offered important opportunities for the public to monitor and challenge logging on national forests. They would become a key tool for grassroots forest protection activists.

The second key difference between the national forest protection efforts and the Headwaters campaign involved the role of the national environmental organizations. Local grassroots activists were able to build the Headwaters campaign largely in the absence of the large national environmental organizations. It was not until the end of that campaign that the Sierra Club national staff got involved in brokering the Headwaters deal over the objections of the grassroots groups. The comparatively minor

Figure 4.1
Volume of trees cut on national forests by year, 1940–2004

role played by the national organizations in that campaign was unusual. The nationals generally played the dominant role in national forest protection issues and, as a result, they could present a significant hindrance to grassroots forest activists who sought greater levels of forest protection than what the nationals advocated.

Nowhere was this dynamic more evident than when some forest protection activists began calling for a complete end to logging on national forests, a goal that became known as "zero cut." The zero-cut activists supported this position with ecological and economic research showing that logging on national forests was ecologically harmful and economically wasteful. Moreover, their research demonstrated that timber cut from national forests was a very small part of the overall wood supply that could readily be replaced by more ecologically responsible alternatives.[2] Above all, the national forests offered a unique space for forests to grow naturally, with trees living for hundreds of years and the forests developing old-growth characteristics that supported native wildlife in a way that the crop-like tree plantations on private timberlands never would. However, the national environmental organizations saw zero cut as too controversial and politically unrealistic. Not only did they not endorse an end to all logging on national forests, but at times they actively undercut the efforts of grassroots forest activists advocating for stronger forest protection measures than the nationals. However, in the late 1980s and 1990s, the dominance of the national environmental organizations was challenged by a plethora of new grassroots forest protection groups, such as Allegheny Defense Project, Buckeye Forest Council, Forest Conservation Council, Forest Guardians, Heartwood, John Muir Project, Native Forest Council, Native Forest Network, Wild Alabama, and many others.

These new groups transformed forest protection activism in two key ways. First, they aggressively applied the National Forest Management Act, the National Environmental Policy Act, and other environmental laws to challenge logging on national forests. At a time when many national environmental organizations focused their resources on Washington, DC, the local grassroots groups were doing much of the on-the-ground monitoring and appeals of the Forest Service's timber sales. This "forest watch" work gave them the legal standing to challenge logging projects in court. Moreover, these grassroots groups were not invested in the insider political strategy of the national environmental organizations and thus were more willing to pursue litigation to shut down logging even when such actions were seen as controversial.

Second, some of these new groups also provided support for grassroots activists who were organizing the members of the Sierra Club to transform that organization from within. These activists believed that zero cut was unlikely to prevail as long as all of the national environmental organizations opposed it. The democratic mechanisms of the Sierra Club gave its members a unique role in shaping that organization's policies. Therefore, zero-cut activists within the Sierra Club formed a dissident network called the Association of Sierra Club Members for Environmental Ethics (later renamed the John Muir Sierrans) to push for stronger environmental protection positions by the Club. The zero-cut activists' attention to transforming a national organization from within distinguished their campaign from the "never mind the nationals" approach of the Headwaters activists. The zero-cut activists' "transform a national" approach was only partially successful, but ultimately grassroots forest protection groups played a central role in dramatically decreasing the logging levels on national forests. The experiences of the zero-cut campaign illustrate both the growing influence of the grassroots groups and the significant hurdles they faced from the national environmental organizations.

Old-Growth Forests and the Northern Spotted Owl

The zero-cut campaign must be situated in the context of the earlier struggles over the old-growth forests in the Pacific Northwest. In the late 1980s and early 1990s, the protection of these forests captured national attention when logging was halted to protect the northern spotted owl, a bird that depends on old-growth forests for its habitat. The history of the campaign for old-growth protection in the Northwest can easily fill a book, so I will not attempt to present a comprehensive account here.[3] Instead, in this section, I focus on those aspects of that campaign that shaped the subsequent zero-cut activism. In part, the success of appeals and litigation in halting logging on national forests throughout the region provided an important inspiration for zero-cut activists. At the same time, the role played by national environmental groups—particularly the Sierra Club—in political deal-making that sacrificed some forests was a source of frustration that would also influence zero-cut activism. I also highlight key contributions made by some of the early grassroots groups that are often overlooked in accounts of the protection of the Northwest forests.

The attention to old-growth forests that emerged in the 1980s marked a notable shift in national forest protection activism. Up until that time, most old-growth protection was an incidental outcome of wilderness protection campaigns. Wilderness offered the primary way to protect a segment of national forest from being logged, although within those campaigns the lower-elevation, old-growth forests were usually traded away for protection of high-elevation rock and ice areas that contained few trees. Through the 1970s and early 1980s, national environmental organizations such as the Sierra Club and The Wilderness Society focused on getting remaining unlogged roadless areas on national forests designated as wilderness. As discussed in chapter 2, the deal-making of these groups resulted in protection of some areas, but it also opened up large areas to logging. In 1984, the wilderness campaign culminated in twenty-two state wilderness bills that covered 8.9 million acres. However, they also released almost 20 million acres of roadless areas from wilderness consideration.[4] After the conclusion of that campaign, the dominant national organizations shifted their attention away from national forest issues.

In their absence, one of the primary advocates for forest protection in the Pacific Northwest was the Oregon Natural Resources Council. This group began as the Oregon Wilderness Coalition (OWC), which formed in 1974 to coordinate the various wilderness protection efforts in the state. However, the organization began to take a different approach when a young biologist named James Montieth became the OWC's coordinator in fall 1974. As a condition of his hire, Montieth insisted that OWC take a position calling for the protection of all of the remaining roadless areas in the region.[5] In responding to doubts about the political feasibility of this goal, Monteith wrote, "Regardless of the so-called realities, it is this goal which will enable us to save maximum wilderness acreage. This goal, by rational perspective, is not unrealistic or unattainable."[6] Nonetheless, this was an unprecedented position, and it put Montieth at odds with national groups such as the Sierra Club. Indeed, in 1978, the Sierra Club tried unsuccessfully to have Montieth ousted.[7]

In the late 1970s, OWC crafted a 325,000-acre wilderness proposal for southern Oregon, but the Sierra Club intervened and cut the proposal back to 102,950 acres.[8] Then in the early 1980s, OWC sought to protect 3.4 million acres of roadless lands in Oregon, but the Sierra Club and The Wilderness Society crafted a proposal for just 1.2 million acres, and ultimately only 950,000 acres were protected.[9]

At the same time that the OWC was working on wilderness legislation, it also began exploring the use of administrative appeals to challenge timber sales on national forests. Reflecting its move away from a singular emphasis on wilderness, OWC changed its name to Oregon Natural Resources Council (ONRC) in 1982. ONRC's campaign to protect unlogged forests outside of designated wilderness areas received a boost when Earth First! got involved in Northwest forest issues. In 1983, Earth First! brought national attention to the fate of the Northwest forests when it staged a series of blockades to prevent the construction of a logging road into the unprotected roadless area of Bald Mountain in southern Oregon. Unlike the national organizations, ONRC did not distance itself from Earth First!. Indeed, the publicity from the Bald Mountain blockade allowed ONRC to raise enough money to file a lawsuit that got an injunction to stop the road construction.

Subsequent Earth First! actions in Oregon focused on other old-growth stands, including a thousand-year-old stand of trees known as "Millennium Grove." It was there that Earth First!ers applied the tactic of tree-sitting for the first time (see chapter 3). While the grove was ultimately cut, there was a proliferation of direct action to protect old growth organized by Earth First! and new spinoff organizations such as the Cathedral Forest Action Group. Their daring actions attracted further media coverage and built public awareness around the issue of old-growth forest protection. Earth First!ers highlight their role in garnering national attention for old growth as one of the group's most notable accomplishments in the 1980s. As journalist Susan Zakin wrote, "In part, it was happening because Earth First! acted without considering the so-called political realities. They fought for the old-growth forests because it was right."[10] While direct action on its own saved few trees, as Zakin noted, "the significance of Earth First! was not in the number of trees it saved but in the debate it provoked."[11]

The old-growth protection efforts received another boost in 1986 when a university student named John Talberth attended a Friends of Cathedral Forest Action Group meeting and proposed to create a timber sale appeal group for them.[12] This entity, known initially as the Citizen's Task Force on Timber Sale Appeals, later became an independent organization and changed its name to the Forest Conservation Council. Talberth's appeals relied on the administrative procedures that were established by the National Forest Management Act, as well as the more general public participation provisions of the National Environmental Policy Act (see appendix). These provisions gave the public the opportunity to challenge the Forest Service's timber sale plans. Talberth's distinct contribution was the tactic of filing multiple timber sales appeals simultaneously on an unprecedented scale. During the 1980s, timber prices dropped and, as a result, timber companies might lose money on the timber sale contracts they had signed earlier with the Forest Service. In response, the agency withdrew all of those timber sales and reissued them at a renegotiated price that would better suit the timber companies. Since the timber sales were reissued en masse, Talberth and the task force responded in kind by appealing almost a thousand timber sales simultaneously.[13] Talberth's appeals provoked outcries from the timber industry's supporters in Congress. And in what would become a recurring motif, his mass appeals were also widely criticized by more moderate groups such as the Sierra Club, which were concerned about a potential political backlash. (One exception was ONRC's director James Montieth, who agreed to sign on to the appeals.) Ultimately about 20 percent of the timber sales were dropped as timber companies were deterred by the prospect of going through the appeals process.[14]

Even when the appeals were not directly successful, they provided the legal standing to then litigate against those timber sales. Talberth's mass appeals thus provided a foundation for subsequent lawsuits. The use of lawsuits was promoted by the Sierra Club Legal Defense Fund, a separate organization from the Sierra Club that took a stronger stance on old-growth protection than the Club. However, the use of litigation faced resistance from moderate environmental groups. As Vic Sher of the Sierra Club Legal Defense Fund explained,

> The environmental community had developed a way of doing business because they weren't used to having access to the courts. They pursued their agenda by trying to maintain access to the congressional delegation. A number of environmental groups tried hard to talk us out of filing the early owl suits. The fear was that they would lose their seat at the table.[15]

Sher became one of the principal architects of a litigation strategy centered around the northern spotted owl, a species that was specifically dependent on old-growth forests as its habitat. Those habitat requirements put the spotted owl directly at odds with logging.

As early as 1972, scientists were sending word to the U.S. Fish and Wildlife Service, Forest Service, and Bureau of Land Management that logging of old growth would imperil the survival of the northern spotted owl. The Fish and Wildlife Service included the owl as a candidate for protection soon after the passage of Endangered Species Act in 1973, but deferred on actually giving the owl the official endangered status that would activate the protections of the law. Likewise, the Forest Service initially rejected requests from owl biologists to not cut old-growth trees in the vicinity of spotted owl nests. Through the 1970s, logging of the old-growth forests continued unabated. In the 1980s, logging on the national forest system surged, with more than 12 billion board feet[16] of timber being cut in peak years and upward of 40 percent of the total volume coming from the national forests of the Pacific Northwest.[17] The old-growth forests of the Northwest were rapidly becoming isolated amid a patchwork of clearcuts. Indeed, in 1992, NASA published satellite photos showing that national forest land in the Pacific Northwest was far more fragmented by logging than the Amazon rainforest.[18]

Amid all of this logging, the northern spotted owl population dropped precipitously. The fate of the owl was tied to the fate of the old-growth forests. It served as the proverbial "canary in the coal mine" for the loss of these forests. Environmental activists saw that the spotted owl would necessarily require protection of extensive areas of old-growth forest, but the Forest Service was resistant to developing an owl protection plan that would protect enough forest from logging. The survival of the spotted owl thus became both a biodiversity protection issue in its own right and also a proxy for the overall condition of the forests.

One of the earliest proponents of incorporating the owl into forest protection strategies was Andy Stahl, a forester in the National Wildlife Federation's Oregon office who was subsequently hired by the Sierra Club Legal Defense Fund in the late 1980s. However, Stahl was also concerned about the political backlash against the owl-based strategy. For example, in 1985, when he learned that two students in California had sent in a petition to list the northern spotted owl as an endangered species, Stahl drove down to California and talked them into withdrawing their petition by arguing that the environmental groups needed to do more preparation before taking that step.[19]

This delay was not due to a lack of scientific evidence documenting the decline of the owl. Indeed, Stahl had located a population genetics expert named Russell Lande whose research concluded that the Forest Service's owl protection program was inadequate and would likely lead to the extinction of the owl. In 1985, Stahl presented Lande's findings to the Forest Service and indicated that he had enough scientific data to shut down logging in the Pacific Northwest. However, Stahl offered to refrain from filing suit while the Forest Service completed a new owl protection plan in exchange for the Forest Service withdrawing six old-growth timber sales. The Forest Service accepted this deal. As Stahl explained, "We knew in 1985 that we could stop every timber sale in old growth and that we could get the owl listed. We decided we shouldn't do it, because public opinion was not developed enough."[20] The pattern of delay by moderate environmentalists continued in 1986 when a gathering of forest activists decided not to petition for listing for the owl despite Lande's findings.[21] During each year that the environmental groups delayed the owl-based litigation strategy, about 4 billion board feet of trees were cut on the national forests of the region and the owl continued to decline.

Then in 1987, a largely unknown grassroots group called Green World forced the hand of the other groups by petitioning the Fish and Wildlife

Service to list the northern spotted owl as endangered. Green World was "a tiny environmental group operating out of a phone booth in Massachusetts," and Stahl was unable to contact the group.[22] Even if he had, he would have been unlikely to talk Green World out of filing the petition. Its founder Max Strahan stated, "We don't cut deals, we won't compromise, because the species are our clients. We have a policy of no extinction. We feel if that line is crossed with any species, you lose the ability to protect life."[23] The other environmental groups had little choice but to join in with their own listing petition for the spotted owl or risk being left out of the listing process. Despite having almost no resources, Green World set in motion developments that would ultimately lead to sweeping changes in the forests of the Pacific Northwest.

At first the Fish and Wildlife Service declined to list the owl in response to these petitions, but litigation from the Sierra Club Legal Defense Fund compelled the agency to reconsider its decision, and the northern spotted owl was finally listed as threatened under the Endangered Species Act in 1990. (This action followed a report by the U.S. General Accounting Office concluding that Fish and Wildlife Service's previous decision not to list the owl had been driven by politics, not science.[24]) While the subsequent debates around owl protection were often framed in terms in the Endangered Species Act, that law actually played a relatively minor role in stopping logging in the Pacific Northwest. Instead, environmentalists found that regulations created through the National Forest Management Act contained a provision requiring the Forest Service to manage its forests in a manner that maintained viable populations of native wildlife. A coalition of environmental groups, with Sierra Club Legal Defense Fund serving as their attorney, filed suit under the National Forest Management Act to challenge the Forest Service's continued logging of owl habitat. The result in May 1989 was a sweeping injunction by U.S. District Judge William Dwyer (a Reagan appointee) that halted new national forest timber sales throughout western Oregon and western Washington until the Forest Service produced a legally adequate plan to protect the spotted owl.

It was an unprecedented victory for environmentalists, but it also produced a strong reaction from the timber industry and its allies in Congress, particularly Senator Mark Hatfield (R-OR). Hatfield was one of the timber industry's staunchest advocates in the Senate, but some national environmental groups were reluctant to risk losing their access to him. For example, Doug Scott, the national conservation director of the Sierra Club, endorsed a strategy based on maintaining this access:

> It requires being in the temple of the enemy. If you're going to have a fight with Mark Hatfield . . . I want to be in the room talking in his ear. He may not vote the way I want but at least I'm talking to him.[25]

As a result, when Hatfield moved to introduce legislation that would overturn Judge Dwyer's injunction, Sierra Club staff worked with Hatfield to craft the bill. The bill included largely symbolic language about the ecological importance of old growth, but the core of the bill was a provision to immediately release 1.1 billion board feet of timber sales from the Dwyer injunction and another provision to require more than 8 billion board feet of timber sales in 1990 that would be preemptively exempted from legal challenges.[26] The legislation was cochampioned by Senator Brock Adams (D-WA), and it was attached as a rider to an existing appropriations bill, which allowed it to avoid going through the full legislative process.

In contrast to the Sierra Club's approach, Senator Patrick Leahy (D-VT) hired an aide specifically to defeat the Hatfield-Adams rider. However, Leahy's efforts were undercut when, at the last minute, the National Audubon Society withdrew its opposition to the rider, and The Wilderness Society also endorsed it (although neither National Audubon Society nor The Wilderness Society were plaintiffs in the lawsuit that the rider would overturn).[27] In October 1989, the rider became law. For grassroots activists in the Northwest, it became known as the "Rider from Hell" because of the devastating logging it produced. Ninety-five percent of the resulting timber sales were in spotted owl habitat.[28] Nonetheless, an Oregon chapter of the Sierra Club gave Hatfield an award for the rider because it included language that acknowledged the ecological importance of old growth.

Due to the opposition from Senator Leahy and others, the Rider from Hell came at a high political cost to its champions. Senator Adams declared that he would not support another such rider, and Hatfield was in a weak position to attempt to do so again. Therefore, when the rider expired at the end of 1990 and the Forest Service began issuing new timber sales that were not exempt from environmental laws, the plaintiffs in the original Dwyer injunction updated their litigation and went back to court. On May 23, 1991, Judge Dwyer granted another injunction that halted new timber sales on national forests throughout the range of the northern spotted owl. This time the injunction was not overturned by rider and it would ultimately stay in effect through 1994 as the Forest Service repeatedly failed to come up with a scientifically credible plan to protect the owl. In issuing his injunction, Dwyer declared,

> More is involved here than a simple failure by an agency to comply with its governing statute. The most recent violation of NFMA [National Forest Management Act] exemplifies a deliberate and systematic refusal by the Forest Service and the FWS [Fish and Wildlife Service] to comply with the laws protecting wildlife. This is not the doing of the scientists, foresters, rangers, and others at the working levels of these agencies. It reflects decisions made by higher authorities in the executive branch of government.[29]

Dwyer's comments reflected the widespread belief that the Bush administration was pursuing a "train wreck" strategy around the spotted owl, avoiding potential resolutions of the issue and instead setting up a crisis in the hopes that it would result in a backlash of public sentiment against endangered species protection.[30] It was expected that this strategy would then lead Congress to rewrite the environmental laws to allow unfettered logging. However, that did not happen, and the net effect of the Bush administration's train wreck approach was to prolong the moratorium of logging in the Northwest for years.

The election of President Bill Clinton in 1992 marked a notable shift in the role of the White House. Clinton campaigned on finding a resolution to the conflict over the Northwest forests. He was elected with strong support from the environmental community, but he also expressed sympathy for the plight of the timber workers affected by the injunction. In contrast to Bush's train wreck strategy, Clinton sought a compromise that would appease both environmentalists and the timber industry. To do so, he established the Forest Ecosystem Management Assessment Team to develop a scientifically credible plan that would allow logging to resume while at the same time protecting the owl and other old-growth-dependent species. The team developed eight proposals. However, it was expected that only two of those proposals would provide sufficient protection for old-growth-dependent species to end the injunction. The most protective proposal would only allow about 200 million board feet of logging per year and the other viable proposal was expected to produce about 1 billion board feet of timber per year. As word of these options leaked out, the timber industry complained that these levels were unacceptable.

In the face of these industry pressures, a subset of the team went back to prepare one more option that sought to offer more timber while still trying to provide credible species protection. This new Option 9 increased the expected annual output to about 1.2 billion board feet, but it did so by allowing logging in forest reserves that would have been protected under the other options, and spotted owl numbers would continue to decline. The Clinton administration championed Option 9, even though the plan was widely criticized by scientists and scientific organizations, including the American Institute of Biological Sciences, the Ecological Society of America, the Wildlife Society, and the Pacific Seabird Group.[31]

The Clinton administration turned to strong-arm tactics to get environmentalists not only to accept Option 9, but also to immediately release millions of board feet of timber sales from their injunction as a good-faith gesture. If the environmentalists did not release those timber sales, the Clinton administration threatened to support legislation that would preemptively exempt logging under Option 9 from litigation.[32] The lawyers for the Sierra Club Legal Defense Fund urged their clients to accept Clinton's demands, but other environmentalists labeled it the "Deal of Shame." Former Oregon Natural Resources Council Executive Director James Montieth argued that "The rider was our opportunity to turn around and face the tiger and fight to win. Most of us felt the rider was a bluff, and if it wasn't a bluff, we thought we could defeat it."[33] Likewise, the new executive director for ONRC Larry Tuttle noted that "at some point we're going to have to have a fight with this administration. If nothing else, that could have revitalized the movement."[34] But the Sierra Club Legal Defense Fund lawyers reportedly threatened to drop ONRC as a client if it refused to accept the deal.[35] Ultimately, the plaintiffs went along with the Deal of Shame and 83 million board feet of forests were released to be cut. As journalist Kathy Durbin noted, "Never before had forest activists been voluntary parties to the destruction of old-growth forests."[36]

Despite Montieth and Tuttle's opposition, ONRC staff member Andy Kerr was an outspoken proponent of the Deal of Shame. Kerr's endorsement of the deal was a reflection of a less confrontational stance spreading among environmental groups in the region. These groups were becoming more dependent on grantmaking foundations. In the early 1990s, foundation funding for forest protection groups in the Northwest grew from 20 percent of their income to 80 percent.[37] Major funders such as Pew Charitable Trusts were not interested in supporting grassroots organizing to challenge Option 9. Instead, they wanted to sponsor monitoring work to ensure the Forest Service complied with Option 9. Larry Tuttle argued that ONRC should not follow Pew's direction. However, Kerr and the ONRC board of directors opposed him and Tuttle left the organization. Kerr then became executive director, and ONRC took Pew's grant. It was left to three small groups with limited funding to challenge Option 9 in court—John Talberth's Forest Conservation Council, Save the West (James Montieth's new group), and Native Forest Council. However, this lawsuit did not succeed. With Option 9 established and the dominant environmental groups in the region tied to maintaining that status quo, forest activism in the Pacific Northwest entered a temporary lull.

Origins of Zero Cut: Protect Our Woods and Heartwood

While the struggles over old growth in the Pacific Northwest propelled the issue of logging on public lands into the national spotlight, the first calls to end *all* logging on national forests did not arise in that region. Instead, they came from an unexpected location—Indiana. As Indiana forest activist Andy Mahler explained, "Nobody was looking at Indiana as being a source of policy leadership on public lands issues. We played that to our advantage. We were the very narrow end of a wedge for the campaign to end logging on national forests."

In the second half of the 1980s, Mahler was the coordinator for Protect Our Woods (POW), which focused on the Hoosier National Forest in southern Indiana. The Hoosier is the second smallest forest in the national forest system. That forestland had been bought back by the federal government as part of a New Deal program. These areas had few old trees or pristine roadless areas, so the protection of old growth and roadless areas was less germane there than in the Northwest. The Hoosier National Forest is also quite fragmented, so many people live adjacent to Forest Service land. When the agency proposed opening up the Hoosier to off-road vehicle use that would be both damaging to the forest and disruptive to those living near the forest, local community members, including Mahler, formed POW and successfully got off-road vehicles completely banned from the Hoosier National Forest in 1987.[38]

Protect Our Woods then turned its attention to the logging taking place on that forest. Initially, the group intended to focus only on the issue of clearcutting, but POW cofounder Bob Klawitter's research found that the logging on the Hoosier was a net money loser for the federal government and largely insignificant in terms of timber supply. Based on this research, POW activist Tom Moore suggested that the organization oppose all logging on the Hoosier National Forest. Although the early group had some loggers as members, POW decided to adopt this position in 1987. The position was further bolstered by the fact that many of the community members in the region had small woodlots on their property. For them, the logging on the national forests represented unfairly federally subsidized competition that drove down the value of their woodlots, so POW's argument against logging was presented not only in terms of the ecological harm from logging, but also in terms of economics. After further research by Mahler, POW decided to extend its position to call for an end to logging throughout the national forest system. Thus, Protect Our Woods became the first organization in the country to endorse the position that was later known as zero cut.[39]

By 1989, the group had helped put enough public pressure on the Hoosier National Forest to bring its logging levels down to almost zero. Mahler then set his sights on expanding POW's model to forest protection efforts for other national forests in the region. To do so, Mahler created a new organization called Heartwood—the name being a conjunction of "heartland" (a nickname for the region) and "hardwood" (the leafy trees that are predominant in that ecosystem). The new group formed in early 1991 around the goals of ending logging on national forests, endorsing sustainable logging practices on private land, and reducing overall demand for wood fiber through conservation and recycling.

At the time of Heartwood's inception, the Fairview timber sale on the Shawnee National Forest in southern Illinois was scene of more than eighty days of blockades by Earth First!ers and other local forest activists.[40] The Forest Service spent $300,000 to bring in law enforcement officers from around the country to police a timber sale that the agency could have bought back from a logging company for $1,100.[41] Fairview was ultimately cut, but Mahler noted that "It was a galvanizing event for Heartwood and kicked us off in a big way." While Earth First! used direct action, Heartwood focused on legal tactics such as timber sale appeals and litigation as a means to stop logging. Mahler noted, "Heartwood existed in parallel with Earth First! and in many ways drew from the same energies but channeled them into different kinds of activities that yielded the desired result."

Heartwood acted as a support network for existing groups that endorsed a zero-cut position, and where no such group could be found, Heartwood helped to form new groups such as the Buckeye Forest Council in Ohio and Kentucky Heartwood. The Heartwood network would ultimately include grassroots groups in sixteen states throughout the Midwest, Great Lakes, and the South, stretching from Pennsylvania (Allegheny Defense Project) to Minnesota (Superior Wilderness Action Network) and down to Alabama (Wild Alabama).

Heartwood had very little funding, so it had to maintain this network over a large geographic area through what Mahler described as a "virtual organization" of people working from their homes (initially connected by phones and fax machines in its pre-Internet phase). To foster personal connections within this network, Heartwood became renowned for its regional gatherings known as "forest councils." As Mahler described, "We always incorporated gatherings in beautiful places, preparing excellent food from local and healthy sources, playing music, making sure that there was always time in the agenda for building the social fabric that is ultimately the key component of building a successful group." In this way, Heartwood fostered its own vibrant movement culture, which was atypical for most grassroots biodiversity groups in the 1990s.

The grassroots groups in the Heartwood network pursued an aggressive campaign of national forest monitoring, timber sale appeals, and litigation, which resulted in widespread stoppages of logging in the region.[42] But politically, Heartwood's opposition to public lands logging faced an uphill battle. As Mahler recalled,

> At the time we started Heartwood, our zero-cut position would not pass what was known as the "laugh test," which means if you were to mention your policy position in a congressional office, the administrative staffer would not be able to restrain a laugh at the notion of trying to accomplish that policy objective, given the political realities of the time. We didn't necessarily think that we would succeed. We all hoped we would. But we didn't do it because we thought we would succeed. We did it because we thought it was the right thing to do and the right position to take. We were going to bring all available resources to bear to change the policy debate and make it something politically realistic in a time when it was completely off the charts in terms of political realism.

In an effort to change those political realities, Heartwood would ultimately find important allies from outside its region, including the Native Forest Council.

Growth of Zero Cut: The Native Forest Council

The Native Forest Council grew out of Tim Hermach's frustrations from trying to work on forest protection within the Sierra Club. Hermach was a former business executive who did volunteer work for the Sierra Club. He became the conservation chair and newsletter editor of the Many Rivers Group of the Sierra Club chapter in Oregon. Hermach's business background led him to become an outspoken critic of the economic irrationality of the Forest Service timber program, but that put him at odds with Doug Scott, the Club's national conservation director, who was concerned about upsetting Oregon's protimber senator, Mark Hatfield. When Hermach published a copy of the newsletter with the headline "Hatfield Misinformed," Scott threatened to disenfranchise the Many Rivers Group.[43] Hermach came to realize that most of his energy was being spent fighting with the Sierra Club's staff rather than with the Forest Service or the timber industry. In 1988, he decided to create an alternate vehicle for his forest protection activism outside of the Sierra Club and thus formed a new grassroots group called the Native Forest Council. (As noted earlier, the Many Rivers Group of the Sierra Club subsequently went on to give Senator Hatfield an award for the Rider from Hell.)

As its name indicates, the Native Forest Council (NFC) was initially focused on an end to all logging of "native forests," that is, old growth and roadless areas. Hermach and NFC proactively challenged the notion that this position was extreme or unreasonable. NFC's educational materials highlighted that in the lower forty-eight states less than 5 percent of the original native forest land remained unlogged. Because most native forests were already cut, NFC contended that anyone who supported continued logging of those forests should be seen as an extremist, rather than those who advocated for full protection of the remaining 5 percent. During the northern spotted owl injunction, NFC became an outspoken critic of proposals that would result in any additional cutting of native forests, and NFC was one of the three groups that filed suit to try to block the Clinton administration's Option 9 plan.

NFC also explored protecting the forests through legislation. In 1989, Hermach developed a proposal for a Native Forest Protection Act, which would have stopped all logging of all native forests on public lands. He traveled to Washington, DC and found fourteen congressional representatives who expressed interest in supporting the bill. However, that initial interest disappeared when those politicians were subsequently visited by lobbyists from the National Audubon Society, National Wildlife Federation, Sierra Club, and The Wilderness Society who did not endorse the NFC bill and discouraged the representatives from doing so.[44]

While the Native Forest Protection Act never became an official bill, it provided inspiration for other environmentalists, and Tim Hermach became one of the forest protection movement's most articulate spokespersons. His core message was that environmental advocates need to ask for what they really want or else they will never get it. His critiques highlighted not only the ways that many environmental groups limited their own advocacy, but also the ways the dominant national environmental organizations sought to constrain grassroots groups from advocating for stronger positions, as evident with their response to the Native Forest Protection Act. Inspired by his own message, Hermach soon became an advocate for ending all logging on national forests. The Native Forest Council became the most prominent national proponent for this goal and popularized the slogan "zero cut" to summarize that position.

Part of the significance of the shift to a call for zero cut was that it fundamentally reframed the terms of the forest debate. Previous advocacy had focused primarily on how much national forest land should be protected as wilderness or debating what level of logging should be allowed, but these goals were not clearly demarcated and thus lent themselves readily to compromise. By comparison, the zero-cut position called the entire Forest Service timber sale program into question. This position put logging advocates on the defensive. It also left little room for the sort of compromises that were the currency of the national organizations that relied on an insider strategy. It is not surprising then that some of the most outspoken criticisms of zero cut came from professional environmental lobbyists rather than from the timber industry.

As with the Native Forest Protection Act, the legislative prospects for zero cut seemed slim as long as they were resisted by the nationals. Therefore, much of the Native Forest Council's energies were channeled into public education. The main vehicle for NFC's outreach was its publication, *Forest Voice*, which was filled with photographs of the vast clearcuts that became a key icon for criticizing public lands logging. Although based in Oregon, the publication allowed the Native Forest Council to reach beyond the Pacific Northwest to a national audience, despite being a small organization. *Forest Voice* also served as a relatively unconstrained space for the community of forest activists to share ideas that were considered too radical for more conventional environmental publications. In this regard, it played a similar role to the *Earth First! Journal*, albeit more focused on forest issues. Above all, *Forest Voice* provided a forum for NFC activists to present sharp critiques when the national environmental organizations promoted compromises that sacrificed some forest protection. For moderate environmentalists, such criticisms seemed divisive, but the confrontational approach of the NFC activists ensured that they would not be ignored.

Reforming the Sierra Club: Origins of the John Muir Sierrans

While Tim Hermach was building an organization for zero-cut advocacy outside of the Sierra Club, other activists worked within the Club to get it to take a stronger stand on forest protection. Two of the principal leaders of the reform movement within the Sierra Club were David Orr and Margaret Hays Young. David Orr had been involved with Earth First! in Texas in the 1980s.[45] At the end of that decade, he moved to California and became active in the Sierra Club as a volunteer. However, he soon grew troubled that the Sierra Club was negotiating an accord with a large Sierra Nevada timber company despite objections from grassroots forest activists. As he talked with other Sierra Club volunteers, Orr heard examples of similar phenomena. As he recalled, "Over and over again the Sierra Club was standing in the way of activism to protect forests. It was a shocking concept to me."

One of the Sierra Club activists with whom Orr talked was Margaret Hays Young, a volunteer leader in the Atlantic Chapter of the Sierra Club in New York. In 1990, the Atlantic Chapter endorsed a proposal by grassroots activists in Montana for comprehensive wilderness protection legislation for their region, which became known as the Northern Rockies Ecosystem Protection Act. However, the Sierra Club national office had not endorsed this legislation. Indeed, in a pattern similar to the Native

Forest Protection Act, after proponents of the legislation found a congressman to introduce it, Sierra Club lobbyist Jim Blomquist reportedly advised the congressman not to do so, and the bill introduction was delayed for two years.[46] Meanwhile, the Sierra Club threatened to suspend the Atlantic Chapter for endorsing the proposed legislation.[47]

In 1991, Orr and Young decided to organize a support network for Sierra Club members who wanted their organization to take stronger stands in its environmental advocacy. In particular, Young suggested that they focus on getting the Sierra Club to endorse an end to all logging on national forests. They called the new group the Association of Sierra Club Members for Environmental Ethics (ASCMEE).[48] The nascent network found support from John Davis, the editor of *Wild Earth*, a conservation magazine started by Dave Foreman and John Davis after they left Earth First!. Through *Wild Earth*, Davis provided a forum for Orr and Young to announce ASCMEE's existence to the larger conservation community and to present their critique of the Sierra Club's policies. As Orr explained in his introductory article,

> The tendency to compromise too much damages not only the environmental legislative agenda, but also activists' morale. Many energetic environmentalists avoid the Club or have left it. We need these knowledgeable people. The common refrains are the Club is out of touch with the grassroots and it compromises too much, too soon. Club leadership generally ignores or dismisses these complaints. In response to these concerns, I am seeking to establish a new organization-within-an-organization, a "fundamentalist" group dedicated to restoring the Club to its rightful place at the cornerstone of the environmental movement, achieving its most noble objective: to protect. Following the spirit of [Sierra Club founder] John Muir, this group will work to make the Club as environmentally ethical, aggressively pro-wilderness, and biocentrically visionary as possible . . . ASCMEE's motto: we're not taking the Club over, we're taking it back.[49]

Then in November 1991, ASCMEE held a protest outside the Sierra Club's national office in San Francisco as the board of directors was meeting. More than 100 people participated in the rally, holding signs such as "Bring Back the Spirit of John Muir" and "Put the Sierra Back in the $ierra Club" and giving speeches criticizing the Club's compromises.[50] The articles and protest attracted some attention, but the ASCMEE founders also received angry letters from people who feared that the reformers' efforts would hurt the Sierra Club. The Club itself threatened legal action against them for trademark infringement because "Sierra Club" was included in their group's name. As a result, ASCMEE went into hiatus in 1992. However, it reemerged in 1993 with a new name—the John Muir Society—in tribute to the Club's founder. The Sierra Club once again threatened legal action after the Club created a new fundraising entity with the same name, so the name was ultimately changed to the John Muir Sierrans (JMS).

Ongoing concerns over the threat of lawsuits from the Club led the John Muir Sierrans to deliberately not create a formal organization. As John Muir Sierran Michael Dorsey explained, "There is no organization— the John Muir Sierrans. It is more of a disposition, a sentiment, a feeling, an orientation, an outlook." JMS provided a shared identity for the reformers, largely maintained through a JMS e-mail discussion list. Through this online community, JMS built a network of grassroots activists to reenergize the democratic mechanisms within the Sierra Club.

A central figure in the development of JMS was Jim Bensman, who had been a volunteer leader in the Illinois Chapter of the Sierra Club since the mid-1980s. In the early 1990s, he also became involved in the growing Heartwood network in his region. While attending a Heartwood forest council gathering and hearing talks on ending public lands logging, Bensman embraced that position. He soon grew frustrated with the Club's refusal to endorse zero cut. The Sierra Club's position was not unique. None of the dominant national environmental organizations advocated zero cut. However, the democratic structure of the Club offered more hope that it might eventually do so. Unlike other national environmental organizations, the Sierra Club's members elected its board of directors, and the Sierra Club's policies could be changed through ballot initiatives voted on by the membership. For the 1994 Club elections, the John Muir Sierrans developed a ballot initiative to have the Sierra Club endorse an end to logging on national forests. Jim Bensman successfully spearheaded the gathering of enough signatures from Club members to get the initiative added to the ballot.

However, the initiative was opposed by the leadership of the Club. Former Sierra Club Executive Director Michael Fischer feared that if the zero-cut ballot initiative were enacted, "the club lobbyists would lose the access [the Sierra Club] has in Washington. Committee staffers would simply refuse to meet with them because their own credibility would be affected if they did. Sierra Club national leaders know that they can't just walk into Congress and say no more clearcutting."[51] When the board of directors met to prepare the election ballot, they modified the ballot language to ask: "Shall the Sierra Club's existing forest policy be retained as is, and not changed by amendment as proposed by the petition?" As a result, members who wanted to support the zero-cut proposal would have to cast a "no" vote on the ballot. The John Muir Sierrans saw this counterintuitive arrangement as an effort by the Club's leadership to thwart the initiative. Bensman tried to send out a message to the Sierra Club's e-mail list alerting them to the confusing ballot language, but the Club responded by threatening to take away his e-mail account.[52]

Ultimately, the 1994 initiative garnered 41 percent of the vote and thus did not pass. The John Muir Sierrans spent the next year trying to negotiate directly with the Sierra Club leadership to endorse zero cut. Those negotiations were ultimately fruitless, but when the John Muir Sierrans returned to try another ballot initiative for the 1996 election, they would find the external conditions more favorable, due to the controversy around a piece of prologging legislation known as the "Salvage Rider."

The Salvage Rider

Between the John Muir Sierrans' first attempt to get the Sierra Club to endorse zero cut and their second ballot initiative, the political conditions changed dramatically. The 1994 congressional elections transformed the political landscape. Republicans won a majority in both houses of Congress for the first time in forty years. The national environmental organizations that had focused on building their ties to Democrats no longer had privileged access to the leaders of the key congressional committees. Instead, those seats were now occupied by conservative Republicans with a stated agenda to roll back environmental laws. Environmentalists in Washington, DC who had relied on an insider strategy watched helplessly as the new Congress introduced a slew of antienvironmental legislation. The primary remaining hope for environmentalists lay in the veto power of President Clinton, but even that was not reliable, as forest activists would learn from the Salvage Rider.

With the Salvage Rider, the timber industry's allies in Congress found a way to once again curtail forest protection activists' enforcement of environmental laws. Logging proponents made use of the publicity around the particularly large season of forest fires in 1994 to argue that the burned areas had to be logged immediately in order to "salvage" as much timber as possible. They claimed that appeals and litigation from forest activists would delay the salvage logging and timber would be lost as dead trees decayed, so that logging should be temporarily exempted from environmental laws in order to expedite the process. Under the proposed legislation, forest activists would have no recourse to the courts to challenge these logging projects on national forests. The provision was written in such a way that, in practice, far more than just dead trees were exempted from legal review. Forest activists vigorously opposed the bill, but environmentalists in DC were not successful in stopping it. The bill was attached as a legislative rider to existing legislation, thus circumventing the full legislative process. As a result, it became known as the Salvage Rider. A rider does not have to address the same issues as the legislation to which it is attached, so the Salvage Rider was attached to the Budget Rescission Act of 1995, which, among other things, included emergency appropriations for the victims of the recent bombing of the federal building in Oklahoma. In June 1995, Clinton cast his first veto ever against this legislation specifically because of the Salvage Rider. However, faced with political pressure to get the appropriation enacted, Clinton eventually signed the bill with the Salvage Rider still included.

The rider covered logging projects involving trees that were labeled as dead, diseased, or dying, a loose category that was applied quite broadly. Moreover, the rider included language that required the Forest Service to re-offer timber sales in Oregon and Washington from 1989 to 1995 that had been subsequently stopped by court injunctions or withdrawn, including some particularly controversial old-growth timber sales created under Senator Hatfield's Rider from Hell. Places that forest activists had thought were safe suddenly were not, and the activists could no longer turn to the courts to intervene. Likewise, political efforts in DC to stop the rider were unsuccessful as environmentalists debated whether to try to overturn all of the provisions of the rider or just reform parts of it. The provisions of the rider would run through the end of 1996, and environmentalists referred to this time period as "logging without laws."

The Salvage Rider had a radicalizing effect on forest activists. With access to the courts blocked and legislative measures thwarted, the locus of activism shifted to the grassroots. Grassroots forest activists found that their sole means to halt the logging was through direct action. After a period of relative quiet during the Dwyer spotted owl injunction and the Clinton forest plan, Earth First! once again became a primary figure in the national forest protection campaign. Even some activists who had previously eschewed Earth First!'s tactics participated in civil disobedience for the first time. For example, during a mass sit-in on the logging road leading to the Sugarloaf timber sale in Oregon, among the ninety people arrested were National Audubon Society vice president Brock Evans and former congressman Jim Jontz, who were sitting with Earth First! cofounder Mike Roselle. (Evans cited the "displeasure and embarrassment I probably caused for the higher-ups" in Audubon from the publicity around the arrest as a factor in being offered early retirement a couple of months later.)

The most successful example of direct action during the Salvage Rider was at Warner Creek in the Willamette National Forest in Oregon. Warner Creek was an unexpected location for success. While many old-growth forest areas scheduled for logging under the Salvage Rider had not been affected by fire, Warner Creek had. An arson-caused fire had opened the previously protected spotted owl habitat in Warner Creek to salvage logging. Prior to the rider, local forest protection activists had gotten a court injunction to stop this logging, arguing that it would harm the forest and ruin an opportunity to demonstrate the benefits of natural forest regeneration following a fire. When their court injunction was overturned by the Salvage Rider in 1995, a small group of Earth First! activists began a blockade of the gravel road to the Warner Creek logging site. They developed a new blockade technology called a "sleeping dragon" in which an activist could chain her wrist to a metal bar within a large block of concrete that was buried in the road. In order to get the activist out of the road, the authorities would need to first dig up the concrete and then cut through it to release her from the sleeping dragon. Faced with this complicated process, the Forest Service withdrew from its first attempt to drive logging equipment up the road to Warner Creek. Buoyed by that success, the blockade grew. Activists developed more elaborate road barriers (including a fort with a moat and drawbridge) and maintained a full-time presence at the site over the weeks and months that followed. This type of long-term, complex blockade became known as a "free state," and that tactic was then adopted by Earth First! activists opposing other Salvage Rider projects. By 1996, there were mass demonstrations, civil disobedience, and blockades occurring throughout the Pacific Northwest. The region saw an unprecedented level of public resistance to logging. Many areas were still cut, but some of the logging projects were stopped. For example, the Warner Creek blockade lasted for a year, until that timber sale was cancelled.[53] Overall, because of the protests, the Salvage Rider only resulted in half as much logging as had been expected.[54]

The ultimate effect of the Salvage Rider was to break down the detente that had existed following the end of Judge Dwyer's spotted owl injunction. It revitalized Earth First!, and the surge in protest brought renewed national attention to the issue of logging on national forests. The controversy also brought embarrassment to the Clinton administration for not vetoing the rider. In 1996, Vice President Al Gore described it as the "biggest mistake we have made."[55] Above all, forest activists were shaken by the loss of places that they had thought were protected. More activists questioned whether any trees could truly be safe as long as logging continued to be allowed on national forests. It was within this radicalized climate that the John Muir Sierrans pursued a second ballot initiative within the Sierra Club to call for an end to the Forest Service's timber sale program.

Chad Hanson and the Second Ballot Initiative

A central figure in second John Muir Sierran ballot initiative and many of the subsequent advances in the campaign to end logging on national forests was Chad Hanson. Hanson became interested in forest protection near the end of college. He was preparing to go to medical school when he joined his brother in 1989 for a multimonth hike along the Pacific Crest Trail, a 2,700-mile route through national forests stretching from the border of Mexico to the border of Canada. It was along that journey that Hanson first encountered large-scale clearcutting. Up to that point, like many people, he assumed that national forests were protected from logging. He was deeply upset by what he saw and resolved to work on ending logging on national forests. After completing the long trek, he became involved with the Los Angeles chapter of the Sierra Club and restarted the chapter's dormant task force on forest issues. He remained involved with the task force even after moving to Oregon to attend law school in 1992. The following spring Hanson became acutely aware of the divergence between his goal of ending logging on national forests and the position of the Sierra Club's leadership. When President Clinton organized a "forest summit" for the Northwest, Hanson got the Los Angeles chapter to sign on to a resolution calling for greater protection for the forests. The media attention that resolution received upset the Club's national leadership. As Hanson recalled,

> From their view, the Sierra Club's name was on record in a very public document which we actually got press over. We were some of the only people from the environmental community that got press during that [forest summit]. We were supporting an end to logging on national forests and that actually made the *Portland Oregonian* at that time. So there was a big firestorm over it and they [the Sierra Club leadership] got really upset. I had a choice at the time. I could either back off or I could push it. I decided to push it and did so for many years afterward. So I quickly became persona non grata in the Sierra Club. I started writing articles for the *Earth First! Journal* about what was going on within the Sierra Club and the failure of the Club to support an end to logging on national forests.

Hanson's actions brought him to the attention of Tim Hermach and the Oregon-based Native Forest Council. NFC provided some support for Hanson's activism, but he still sought to work within the Club. In particular, his analysis of the history of environmental legislation found that no major law had passed without the endorsement of the Sierra Club. As Hanson explained, "Even if I wanted to cancel my membership and say, 'To heck with the Sierra Club; I'm going to do my own thing,' the Sierra Club would always be there right in the middle of the path like the proverbial 800-pound gorilla, blocking the way. There was no way to get around it." His articles in the *Earth First! Journal* led David Orr to contact him and bring him into the John Muir Sierrans. There he helped gather signatures for the ill-fated first JMS ballot initiative. Then in 1995, when the John Muir Sierrans decided to try a second ballot initiative for the 1996 Sierra Club elections, Hanson took on the role of coordinating the effort.

A key concern for Hanson was how to avoid a repeat of the situation where the Sierra Club board reframed the initiative language for the ballot such that members would have to vote no in order support an end to national forest logging. Examining the Club's bylaws, Hanson found that if he boiled the initiative down to a one-sentence question, opponents within the Club leadership would have no latitude to change that language when preparing the ballot. Hanson's initiative simply asked, "Shall the Sierra Club support protecting all federal publicly owned lands in United States by advocating an end to all commercial logging on these lands?"

With the ballot language resolved, the John Muir Sierrans would have a few months during the summer to gather the 2,000 signatures from Club members required to get the initiative on the ballot for the following spring. JMS provided a network of eager volunteers to undertake this grassroots organizing, but it was no easy task to reach large numbers of Club members through face-to-face contact rather than the expensive mass mailings used by the Club itself. Hanson's mother and sister took an active role in gathering signatures in Los Angeles. Los Angeles was home to the largest Sierra Club chapter, and many members were quite active in the weekend hikes that the Club organized in the four national forests surrounding the city. With petitions in hand, Hanson family members would wait at the trailheads on weekend mornings for the clusters of members embarking on their hikes.[56]

That summer, Chad Hanson had graduated from law school and moved to New York City to live with his girlfriend. However, three and a half weeks before all of the petitions were due, their relationship ended and he moved out of her apartment. Hanson coordinated the final stages of the nationwide signature-gathering from a payphone in a subway station every evening, as the tiny hotel room into which he had moved had no phone. Collecting the signed petitions would prove even more difficult than getting the signatures. The petitions that had been sent to volunteers throughout the country had been printed with his girlfriend's Brooklyn apartment as the return address. But she would not give him newly arriving petitions, and he had not yet received enough signatures to qualify the initiative. Therefore, he decided that he needed to get the petitions directly from the mail carrier before they were delivered. That would not be an easy task. Hanson was working as a high school teacher in Queens. He would depart school immediately after the end of class and catch the subway to Brooklyn. But the subway route that first took him through Manhattan was u-shaped and thus added on so much time that he kept missing the mail carrier. Hanson then determined that he could circumvent the u-route and cut forty minutes off his travel time if he sprinted through a rough neighborhood in Queens and caught the train on the other side. It worked. As Hanson recounted,

> I caught up with the mail carrier outside the apartment before he
> dropped off the mail. I said, "Here's the deal. I've got a really, really
> weird story for you. I promise you I'm legit and the fate of our national

forests could be in your hands. Here's the story. Can you sit down for five minutes because I just need to tell you this? My ex-girlfriend's in that apartment. She doesn't want to talk to me. She's not going to give me my mail. I need this stuff. It's important mail. They are the signatures I need to qualify this initiative in the Sierra Club in three weeks. So what I'm asking you is—every day I'm going to take the subway, then sprint across Queens to pick up this other train so I can get here in time. I'm going to run here. I'll be coming up sweating and panting, so don't be alarmed. I'm going to intercept you just before you get to this apartment and if you could, just out of the goodness of your heart—because I know this is against policy—just set aside the stack of petitions for me, put a rubber band around it, and just take it out of your bag and hand it to me everyday." He said, "I can't do that. It's against policy. I could get fired." And I said, "I've got no options here. I'm just asking you, please." And he said, "Okay."

The effort paid off. The initiative qualified by a margin of only fifty signatures.

Once the initiative had gained enough signatures and was on the ballot, the John Muir Sierrans had ten weeks to rally support among Sierra Club members before the voting period ended. They began sending letters to the Club's volunteer leaders in each chapter, using the addresses listed on the back of each chapter's newsletter. However, they were immediately contacted by the Club's leadership and told that those names and addresses were proprietary information of the Club and that use of that information would cause their initiative to be disqualified. They were also unable to use the Club's e-mail system or phone lists. With all other avenues blocked, Hanson and Orr decided that their only option for reaching Club members was to directly attend chapter meetings throughout the country. For the next two months they drove from chapter to chapter speaking to the membership directly, at times driving hundreds of miles per day. They did this as volunteers. At each stop, they relied on the donations they received to help pay for the gas to get them to the next location. After the meetings, they slept on people's floors and hoped that someone would buy them dinner. They encountered some opposition along the way, but ultimately this grassroots outreach was successful. On April 21, 1996— the 158th anniversary of John Muir's birth—the Sierra Club announced its election results. The JMS initiative had won, receiving the support of two-thirds of the Sierra Club's membership. The Sierra Club became the first national environmental organization to endorse an end to commercial logging on national forests.

John Muir Project and the National Forest Protection and Restoration Act

When the Sierra Club officially endorsed the zero-cut position, that was just one step, albeit an important one, toward achieving an end to logging on national forests.[57] Much work remained to get the Club to implement this policy, and the John Muir Sierrans needed to find a way to sustain their activism. They were still encountering strong resistance to the no-logging position among the Sierra Club staff, so John Muir Sierrans were unlikely to find employment within the Club. Moreover, being on staff would have constrained their ability to criticize Club policies, so they had to make their livelihood outside of the organization. Many John Muir Sierrans worked day jobs unconnected to environmentalism and did their forest protection advocacy as volunteers in whatever spare time they could find. However, with much of that spare time spent in exhausting struggles with the Sierra Club staff and leadership over the implementation of stronger forest protection policies, this path could easily lead to burnout.

The proliferation of new grassroots forest protection groups provided an important alternative for some John Muir Sierrans. Employment within these groups allowed the activists to work full time on forest protection while continuing to do their JMS work as volunteers within the Club. For example, Jim Bensman, architect of JMS's first ballot initiative, was hired as the forest watch coordinator for Heartwood. Chad Hanson and David Orr pursued a similar course. In fall 1996, they began their own grassroots forest protection organization, which they named the John Muir Project, under the fiscal sponsorship of Earth Island Institute. Former Sierra Club Executive Director David Brower founded Earth Island Institute in 1982 to serve as an umbrella organization for innovative new grassroots environmental projects.[58] It provided tax-exempt status and administrative support for its member groups, as well as publicizing their activities through a shared publication, *Earth Island Journal*. The John Muir Project found support from a handful of sympathetic foundations and donors, with the Foundation for Deep Ecology, Levinson Foundation, Patagonia, and Turner Foundation as some of its earliest funders. These resources allowed Hanson and Orr to pay themselves modest salaries and eventually add a small staff.

Despite its small size, the John Muir Project sought to maximize its influence by providing resources to encourage and assist other organizations in advocating for zero cut. One of the first and most influential materials produced by the John Muir Project was a report by Chad Hanson titled, "Ending Timber Sales on National Forests: The Facts." The report was the result of a year of research into the economic costs of national forest logging. Previously, other environmental groups had identified that the Forest Service sometimes spent more money preparing a timber sale and building the logging roads than it received in payment from the timber company, but Hanson's research went beyond individual below-cost timber sales to assess the balance sheet of the entire Forest Service timber sales program. Hanson identified an unprecedented amount of expenses. The Forest Service's timber sale program as a whole was a money loser, operating at a net cost to taxpayers of at least $791 million in 1996. (Later analyses found that the timber program lost more than $1 billion per year in subsequent years.[59]) Hanson's findings were verified by the Congressional Research Service, the nonpartisan research arm of Congress.[60]

Not only did the John Muir Project report present the Forest Service timber sales program as a form of taxpayer-funded subsidy to the timber industry, but it also questioned the need for that logging. For example, logging on national forests was often justified in terms of the employment it provided for timber workers. However, the report determined that the number of people employed by logging on national forests was so small compared to the total annual cost of the timber sale program that if the funding spent subsidizing the timber program was instead redirected to employ all the public lands timber workers to do ecological restoration rather than commercial logging at the same pay rate, that program would cost taxpayers hundreds of millions of dollars less per year than the timber sale program. In terms of wood supply, the report found that less than 4 percent of U.S. wood consumption was provided by national forests, an amount that could be readily replaced by a bit more recycling or alternative materials. In sum, the report portrayed the Forest Service's timber program not only as ecologically harmful, but also unnecessary and economically irrational.

By critiquing the timber sale program as a whole, the John Muir Project report helped to shift the debate from the question of how much logging there should be on national forests to the question of whether there should be a timber sale program at all. In so doing, it preemptively challenged the logic of political bargaining on the issue. When some environmentalists argued that there should be *x* amount of logging and the timber industry argued that there should be *y* amount, some quantity between *x* and *y* might be treated as a reasonable compromise. However if the entire national forest timber program was irrational, then no quantity of logging made sense.

The economic dimensions of the John Muir Project report also proved to be a helpful resource in garnering political support as the campaign to end logging on national forests moved into realm of legislation. Soon after the Sierra Club ballot initiative passed, Hanson traveled to Washington, DC to find a champion for zero-cut legislation. He was concerned by rumors that the Sierra Club's staff did not plan to actively implement the Club's new position. Indeed, when he arrived in the Club's DC office he received a chilly reception from most of the staff. He was told that pursuing zero-cut legislation was politically unrealistic, that he would never find a member of Congress willing to support it, and that he would be laughed out of every congressional office for even suggesting it.[61]

Hanson was joined in DC by Rene Voss, a Sierra Club volunteer from Georgia who had been very active in gathering signatures for the ballot initiative. Voss arranged a meeting with Rep. Cynthia McKinney, a maverick African American Democrat from Georgia who had recently been elected with help from the local Sierra Club activists. Within the first minute of talking with Hanson and Voss, McKinney offered to be the champion for legislation to end logging on national forests.[62] Over the following year, they worked with McKinney's staff to craft the legislation.

The bill was titled the National Forest Protection and Restoration Act (NFPRA). While previous forest protection bills were filled with pages of legislative language creating extensive regulations and oversight procedures for logging, NFPRA succinctly ended the Forest Service timber sales program, so the bulk of the new bill was dedicated to determining how to redistribute all of the money saved by this action. Up to a third of the annual savings could be reappropriated to fund ecological restoration on public lands, which would provide jobs in the woods for displaced timber workers to repair the damage that had been caused by the logging program. The bill also sent funds to the Environmental Protection Agency to develop and promote "tree-free" alternatives to wood and paper as a way to reduce demand for timber and mitigate the small reduction in wood supply resulting from the legislation.[63] Finally, after those expenditures, the remaining hundreds of millions of dollars saved each year from ending the timber sale program would be sent back to the U.S. Treasury to reduce the federal deficit. Thus NFPRA not only offered strong forest protection, but also appealed to fiscal conservatives. As a result, zero-cut activists found a Republican congressman from Iowa, Rep. Jim Leach, who agreed to serve as cochampion for the legislation, thereby giving the zero-cut bill bipartisan support at a time when Republican support for forest protection legislation was rare.

The zero-cut activists recognized that NFRPA would not quickly become law. Indeed, most notable public lands protection laws had taken many years to achieve passage. For example, the Wilderness Act of 1964 had first been introduced in Congress in 1956. More recently, the California Desert Protection Act of 1994 had taken ten years to become law. And both of these laws affected only a small fraction of public lands compared to NFPRA. Congressional opponents of public lands protection were often well positioned to prevent environmental legislation from moving out of committee for a vote by the full Congress. The zero-cut activists' legislative strategy therefore focused on finding more cosponsors for NFPRA. Cosponsorship was an official endorsement of the legislation by other representatives. As more added their names to the bill, that put pressure on the committees to allow the bill to move forward.

NFPRA was officially introduced in October 1997 with nine original cosponsors. In January 1998, Rene Voss joined the staff of John Muir Project to provide a full-time presence in Washington, DC. (The rest of the staff was based in Los Angeles, which has the highest concentration of congressional districts in the country.) Voss's primary role there was not lobbying politicians. Instead, as he explained, much of his time was spent "monitoring the national environmental groups and trying to keep their message consistent with ours so there weren't statements made by the other groups saying, 'No, we oppose this legislation,' or they weren't doing something contrary that would promote logging, as had happened in the past." (While none of the national environmental organizations other than the Sierra Club endorsed zero cut, Voss noted that some staff members of those groups confided in him that they personally supported an end to logging on national forests even though their organizations would not officially embrace that position.)

Meanwhile, much of the effort to recruit cosponsors for NFPRA took place outside of DC through grassroots organizing at the local level. By the end of that congressional session, the list of cosponsors had grown to thirty-four. The bill was reintroduced in 1999 and reached ninety-nine cosponsors in that session. It then passed the 100 cosponsor mark after it was reintroduced again in 2001.

While zero-cut activists were engaged in a legislative campaign, they did not follow an insider approach to electoral politics. They advocated for an end to logging on national forests at a time when the professional environmental lobbyists treated that position as politically unrealistic. And unlike the professional lobbyists, they did not rely on privileged access to politicians to achieve their goals. Instead, they saw their power outside of Washington in district-level grassroots political organizing, as well as lit-igation to directly block logging, as I discuss later.

The Sierra Club Board of Directors

Although the Sierra Club staff was no longer able to publicly criticize the zero-cut position, the John Muir Sierrans believed that the staff did not intend to put much effort into promoting that position either. Many John Muir Sierrans saw Sierra Club executive director Carl Pope as the prin-cipal impediment. As Chad Hanson described,

> Carl has a totally different vision of how to achieve social change for the environment from the vision I have and other John Muir Sier-rans have. In Carl's world, things are accomplished through an insider's approach. They are accomplished through cultivation of relationships with a particular political party. You choose sides and you stick with that side. And the side that he chose and stuck with was the Democratic Party. [From Pope's perspective,] we had better chances of protecting the environment by siding with the Democ-rats, even to the extent of looking the other way when they do the wrong thing or when they refuse to support us in doing the right thing.

The John Muir Sierrans felt that they would have to challenge Pope's administration in order to ensure that the Club played more than just a perfunctory role in the campaign to stop logging on national forests. Over-sight of the executive director lay in the hands of the Club's board of direc-tors. However, in the early 1990s, the John Muir Sierrans saw the board as being in the hands of an "old guard" who would not challenge Pope's approach. In Hanson's analysis, "There was always this restraint on the part of the board that's allowed Carl to operate in ways that fundamentally perpetuated this dynamic."

The democratic structure of the Club offered the John Muir Sierrans the opportunity to change the board. Unlike most other national environmental organizations, the Club's board of directors is directly elected by the membership. Five of the fifteen board members are elected each year for three-year terms. An official nominating committee selects the board candidates, and traditionally those selections had maintained the old guard. However, Club members could add additional board candidates by petition if they collected enough signatures from other members.[64] The John Muir Sierrans provided a network of volunteers to gather signatures for reform candidates.

In 1994, Laura Hoehn became the first JMS candidate elected, though she faced the isolation that came with being the sole John Muir Sierran on the board. Then in 1995, former Sierra Club Executive Director David Brower agreed to run for the board as a JMS candidate. Brower was one of the best-known environmentalists in the world for his role in revitalizing the activism of the Sierra Club when he was its first executive director in the 1950s and 1960s. However, he had been forced out by the Club's board in 1969 (see chapter 2). Since that time, Brower had become an outspoken critic of compromise in environmental advocacy. In a widely circulated letter rebuking the strategy of Sierra Club National Conservation Director Doug Scott, Brower wrote,

> My thesis is that compromise is often necessary but that it ought not originate with the Sierra Club. We are to hold fast to what we believe is right, fight for it, and find allies and adduce all possible arguments for our cause. If we cannot find enough vigor in us or them to win, then let someone else propose the compromise. We thereupon work hard to coax it our way. We become a nucleus around which the strongest force can build and function.[65]

At the age of eighty-three, Brower ran for the Club's board of directors as a JMS candidate in a renewed effort to implement that approach, and he was elected.[66]

In 1996, Susan Holmes was the next person elected to the board with JMS support, but she resigned her directorship early when she was offered a staff position. In 1997, JMS presence on the board grew dramatically as three of the five new board members were JMS petition candidates. With the successful introduction of the National Forest Protection and Restoration Act under his belt, Chad Hanson was elected. He was joined by Betsy Gaines, a wilderness advocate from Montana, and Michael Dorsey, an environmental justice activist who became the first African American man to serve on the Club's board.[67]

The influx of John Muir Sierrans shifted the balance of power on the board. The reformers now represented a sizable voting block that could not easily be ignored. The 1998 election brought a new slate of JMS candidates, but they faced new obstacles around their ballot statements. As candidates, they were allotted space within the Club's election materials to make a statement. The John Muir Sierrans used their spaces to raise criticisms about the direction and management of the Club. The Club responded by deleting parts of JMS candidates' statements and inserting rebutting text in bold letters beneath their statements. Hanson investigated what he saw as illegal actions by the Club around the ballot. Initially, he was denied access to the relevant Club documents even though he was a member of the board of directors. Ultimately, he had to get a lawyer and threaten to sue the Club until the JMS ballot statements were returned largely to their original form.[68] That year, another three JMS candidates won—Veronica Eady, an environmental justice lawyer and the second African American woman ever elected to the board; Jennifer Ferenstein, another Montana wilderness advocate; and David Brower was reelected.

With six John Muir Sierrans on the board, they only needed two more seats in the 1999 election to gain a majority. At that point, the John Muir Sierrans had also developed a presence on the Club committee that nominated candidates for the board. As a result, the two JMS-endorsed candidates who won seats in 1999—Rene Voss of John Muir Project and Charlie Ogle of the Constitutional Law Foundation—did not have to be added to the ballot by petition. The addition of Voss and Ogle meant that within five years the John Muir Sierrans had gone from having one lone board member to holding a majority.

With a JMS majority, there was talk that Executive Director Carl Pope might be ousted in the coming term.[69] The direct oversight of Pope lay with the board president. David Brower expressed an interest in serving as president as the capstone to a lifetime of involvement with the Sierra Club. However, by the time of the board meeting to elect the next president, Ferenstein, Gaines, and Ogle had publicly distanced themselves from JMS, and Brower's bid was defeated, with a non-JMS candidate instead becoming board president. Critics of Ferenstein, Gaines, and Ogle's action suggested that they had been co-opted by Carl Pope and the old guard.[70] (Ferenstein was elected to the executive committee of the board in 1999 and subsequently became board president in 2001.) However, Ferenstein asserted that in running as a JMS candidate, "I never promised them that I would always vote a certain way." While Ferenstein saw her actions as indicative of her independence, other John Muir Sierrans saw her actions as indicative of a key structural problem within their network. JMS's lack of formal organization meant that it had no concrete mechanisms to keep its candidates responsible to the larger group once they were elected. As David Orr lamented, "How do we do an insurgency when we have no way to hold people accountable?"

Ultimately, the 1999 elections were the high-water mark for the John Muir Sierrans' presence on the board. Veronica Eady resigned in 1999, citing health problems. David Brower resigned in protest over what he saw as inaction by the Club in early 2000, just six months before his death. While Dorsey and Hanson were reelected that year, there was still a net loss in JMS numbers on the board, and the number of JMS candidates elected would continue to drop from then on.

Nonetheless, by that time, JMS had enacted some significant changes within the Club. Most notably, they spearheaded a new national priority campaign program within the organization whereby the Club would allocate substantial funding to the environmental issues that received the most votes from the Club's membership. This program bolstered the Club's environmental advocacy and gave the membership more control over the direction of the organization. For the campaign to end commercial logging on national forests, these developments brought a notable increase in resources. In 1997, the Club allocated less than $15,000 for that campaign; by 2000, it had a budget of more than a quarter of a million dollars.[71]

Beyond the Sierra Club: Timber Sale Litigation and the National Forest Protection Alliance

While the John Muir Sierrans were working to transform the Sierra Club's position on zero cut, much of the work of directly stopping logging on national forests was being done by a loose network of grassroots forest protection groups. These groups used the provisions the National Forest Management Act and the National Environmental Protection Act to monitor and appeal the Forest Service's proposed timber sales as well as the larger forest-wide management plans. At times, their appeals were enough to get timber sales withdrawn. Other times, if the appeals were rejected by the Forest Service, the forest protection groups would turn to litigation to stop the logging. Challenging individual timber sales helped to lower overall logging levels, but the most dramatic impacts came from the lawsuits that halted logging throughout an entire national forest or even an entire region. The northern spotted owl injunction described earlier was the most famous example, but it was far from the only one. By the end of the 1990s, lawsuits had resulted in at least temporary shutdowns of various national forests throughout the country. For example, John Talberth of the Forest Conservation Council—whose mass appeals had provided an impetus for the spotted owl injunction in the Pacific Northwest in 1989—moved from Oregon to New Mexico in the early 1990s and began applying his timber sale appeal and litigation tactics there. During the 1990s, the Forest Conservation Council temporarily merged with another grassroots group called Forest Guardians and together they were part of a coalition of groups that shut down logging on the eleven national forests of the Southwest for sixteen months (see chapter 5). Likewise, by the time that Andy Mahler stepped down as executive director in 1999, Heartwood and its affiliated groups had halted logging on eight national forests in the Midwest, East, and South.[72]

While Heartwood began with a stated goal of ending all national forest logging, some grassroots forest groups initially did not explicitly call for zero cut, even though they were involved in stopping much logging. Some of those groups were concerned that the major environmental grant-making foundations would not fund them if they officially endorsed zero cut. In particular, grassroots activists highlighted Pew Charitable Trusts as a prominent funder that would not support zero-cut groups, pointing to its role in an aborted national forest campaign coalition. Journalist Mark Dowie reported, "As they dangled millions of dollars in front of starving organizations, the funding group led by Josh Reichert [of Pew] stipulated that no one advocating zero cut, criticizing corporations by name, or producing ads that did so would be eligible for membership in the forest coalition—or for funding."[73]

However, zero-cut groups were able to find a handful of supportive funders. The largest was the Turner Foundation. Others were small to mid-sized grantmakers such as the Foundation for Deep Ecology, Levinson Foundation, and Patagonia. A common theme among these funders was that their program staff came from activist backgrounds in some of the more radical environmental groups: Peter Bahouth of the Turner Foundation had been executive director of Greenpeace USA; Bill Devall and John Davis of the Foundation for Deep Ecology and Charlotte Talberth of Levinson Foundation had been involved with Earth First! in the 1980s; and John Sterling of Patagonia had previously worked for Earth Island Institute.

By the second half of the 1990s, more and more grassroots forest protection groups endorsed zero cut. The experience of the Salvage Rider had galvanized them. As Matthew Koehler of the Montana-based Native Forest Network described, "Coming out of the Salvage Rider, there was a lot of energy around the idea of just ending all the logging on national forests." Moreover, the Sierra Club's endorsement of zero cut and the subsequent introduction of the National Forest Protection and Restoration Act made the zero-cut goal seem more feasible.

In 1996, the growing network of groups seeking to end national forest logging began to create a more formal coalition, initially known as the Zero Cut Campaign and later becoming the National Forest Protection Alliance. In part, the coalition was motivated by a desire to increase coordination among these groups in support of the legislation. The coalition also provided greater coordination for litigation. In 1997, John Talberth and Bryan Bird of Forest Conservation Council/Forest Guardians initiated a campaign to appeal all of the Forest Service timber sales in the country, and the National Forest Protection Alliance and its member groups joined in this nationwide mass appeal process. At the same time, the coalition gave national visibility to small zero-cut groups that had previously been consigned to the margins of the forest protection movement.[74] By 2000, the National Forest Protection Alliance represented eighty-nine member groups, and 200 organizations had endorsed the National Forest Protection and Restoration Act.

Funding was another consideration in the development of National Forest Protection Alliance. Similar to the Headwaters Forest Coordinating Committee (see chapter 3), there was hope that a formal coalition organization might attract new funding. The coalition initially spent a sizable portion of its budget on the services of outside consultants in the hopes that a more professional image would appeal to the dominant environmental grantmaking foundations.[75] Yet despite the efforts of the coalition and the accomplishments of its member groups, zero-cut groups were not able to attract notable support from those foundations. Instead, those funders soon poured their resources into the Clinton administration's "Roadless Rule."

The Roadless Rule and the Heritage Forests Campaign

The Roadless Area Conservation Rule was the hallmark forest policy measure of President Clinton's second term. It attracted much media attention and received extensive praise from moderate environmental organizations. However, it simultaneously drew energy away from the zero-cut campaign at the point that the campaign had started to make notable headway.

Announced in October 1999, Clinton's initiative proposed to end construction of new roads in areas of national forest larger than 5,000 acres that did not already contain roads. The plan as it was proposed would affect more than 40 million acres of federal land. Since logging operations generally relied on roads to remove the cut trees, those roadless areas were expected to be relatively safe from logging. Clinton's announcement began a year-long public comment process to assess the proposal. At the same time, some of the national and regional environmental organizations that had declined to support zero cut mobilized a massive campaign to generate support for roadless area protection under the auspices of a new entity initiated by the Pew Charitable Trusts called the Heritage Forests Campaign (HFC). (Ironically, some of these organizations had at one time dismissed James Montieth's call for protection of all roadless areas as extreme and politically unrealistic.) Funding poured into the HFC, with Pew alone providing more than $3.5 million.[76] In appealing to funders, the HFC pointed to its polling figures showing 65 percent public support for roadless area protection. However, the same pollsters had previously found higher public support for ending all logging throughout the national forests, with about 69 percent in favor of this stronger position.[77]

In many ways, the Heritage Forests Campaign was the opposite of the zero-cut campaign. Zero cut was a grassroots campaign that sought to transform political realities by empowering activists with new skills and aggressive strategies. In contrast, the roadless campaign had a top-down structure. The dominant environmental organizations primarily acted to defend the Clinton administration's position. HFC's demands went only slightly beyond the initial proposal by calling for inclusion of the Tongass National Forest of Alaska. And the HFC used its extensive funding in ways that did little to empower the public. Instead, public participation was largely confined to sending off preprinted postcards and e-mails in support of the HFC's position.

That said, the HFC avoided some of the problems that accompanied earlier forest protection efforts by national environmental organizations. In particular, it did not directly sacrifice some environmental protections in exchange for others, as had happened with the "release" provisions of some wilderness bills or the Endangered Species Act exemptions that were tied to the Headwaters deal (see chapter 3). Nonetheless, the HFC operated in a manner that was ultimately detrimental to zero cut. At the point when zero-cut advocates had been making advances in raising the larger question of whether there should be a Forest Service timber sale program, the HFC shifted the terms of the forest debate back to more limited questions of whether there should be logging on particular parts of national forests—Should roadless areas be protected? What size roadless areas should be included? And should some roadless areas such as the Tongass be excluded from protection? The protection of the rest of the national forest lands disappeared from this discussion.

Another problem with the roadless campaign was that it overstated the benefits of the initiative for forest protection. In announcing his roadless areas proposal, President Clinton noted that less than 5 percent of trees cut in the United States came from national forests and that less than 5 percent of the logging on national forests took place in roadless areas; he added, "We can easily adjust our federal timber program to replace 5 percent of 5 percent."[78] Despite the large acreage involved, the Clinton roadless areas plan would have little or no impact on actual logging levels. (By comparison, appeals and litigation by grassroots forest activists had contributed to a 78 percent drop in national forest logging levels between 1988 and 1999. See figure 4.1.) Yet following Clinton's announcement, the news coverage was filled with unbridled acclaim from the national environmental organizations. As the *New York Times* wrote, "Environmentalists hailed the order as rivaling only the steps taken by President Theodore Roosevelt in laying the foundation for today's national forest system."[79] Such effusive praise for Clinton's measure made calls for additional forest protection beyond those roadless areas seem excessive and unreasonable.[80]

This is not to say that the environmental community could not have supported roadless protection without undermining zero cut. The HFC potentially could have welcomed the increased protection offered by the Roadless Rule while at the same time using the attention generated by the issue to highlight the need for a complete end to national forest logging. That approach might have upset the administration by presenting Clinton's proposal as insufficient, but it would have provided a better groundwork for activists to pursue greater forest protection after the Roadless Rule had been enacted. The Heritage Forests Campaign thus passed up an important opportunity to advance the larger forest protection goals in the long term, while at the same time securing incremental gains in the short term.

Some National Forest Protection Alliance groups tried to integrate a zero-cut message while working for enactment of the roadless initiative, but those efforts bore little fruit for the zero-cut campaign. With funding concentrated in the hands of large national environmental organizations at the top of the HFC, the zero-cut groups had little chance to advance a stronger message amid the roadless debate. Moreover, other National Forest Protection Alliance members saw those attempts as an unnecessary drain on their coalition's limited resources. As John Talberth, the first board president of the alliance, argued,

> I think those groups should have realized the roadless campaign was covered. All of the national groups were putting all of their energy into that. The niche that we were filling was the end commercial logging campaign and we should have not gotten involved in that roadless thing. We should have continued to focus on our mission. Instead, we just became one of the many groups working on the roadless campaign. It was a conversion that alienated a core part of the group.

Moreover, for those zero-cut activists who thought that the roadless campaign could be a brief hiatus before returning to their larger goal, that hope would prove illusory. Immediately upon taking office at end of January 2001, President George W. Bush suspended Clinton's Roadless Rule. That rule would remain mired in a limbo of litigation, policy revisions, and public comment processes that would continue to occupy the time and resources of environmentalists throughout Bush's presidency.

Fire and Funding during the George W. Bush Years

The setbacks for forest protection under the George W. Bush administration were not limited to the rollback of the Roadless Rule. The new administration was also quite proficient at using concerns over forest fire as a pretext to reduce national forest protection, although this approach did not originate with Bush. It had been promoted by the timber industry throughout the 1990s when their job-based justifications for more logging became less credible as it was revealed that timber industry job loss resulted largely from the timber companies' increasing mechanization, their depletion of forests, and their export of unprocessed logs overseas.[81] The "forest health" argument took a different angle, asserting that logging was needed to "protect" forests from fire. However, this claim was fraught with contradictions. Fire is a natural and necessary part of most western forest ecosystems. Moreover, fire is underrepresented in these forests compared to natural levels due to decades of fire suppression.[82] Nonetheless, the idea of logging for forest health gained credence by drawing on the claims of some scientists that, in circumstances where fire suppression had disrupted the cycles of forest fire, restoration work to reduce the resulting brush and small trees could sometimes help to restore more natural fire patterns. However, the material that needed to be cut for this restoration work was generally too small to be used for lumber and thus had little or no commercial value. By contrast, the more commercially valuable large trees favored by the timber industry were also the most fire resistant and did not need to be removed. Indeed, logging the larger trees whose branches shade the forest floor could leave the remaining forest hotter, drier, and more fire-prone.[83] In spite of these contradictions, logging advocates sought to conflate commercial logging and ecological restoration. By the end of the 1990s, many of the largest logging projects on national forests were being justified under the pretext of forest health.[84]

The justification of logging in terms of forest health gained ground in part because of divisions between zero-cut and non-zero-cut forest activists over how to respond to it. Most zero-cut groups did not oppose genuine ecological restoration activities that involved the cutting of brush and smaller trees. However, they believed that the only way to ensure that ecological restoration projects were not turned into logging-as-usual under a new name was to remove the commercial incentive. If the Forest Service was not permitted to sell the material that was cut during restoration projects, it would have no incentive to promote logging of larger trees that did not need to be cut for restoration purposes. Accordingly, from the mid-1990s onward, zero-cut activists increasingly adopted the phrase "end commercial logging" to more accurately describe their position. This position was reflected in the provisions of the National Forest Protection and Restoration Act, which would end commercial logging on national forests but included no restriction on noncommercial, science-based ecological restoration activities. In this way, zero-cut activists offered a comprehensive alternative to so-called forest health logging.

By comparison, environmental organizations that did not endorse zero cut were more constrained in rebutting the idea of forest health logging. Because they did not want to be seen as opposing all commercial logging, they needed to find some logging projects to support and thus were more likely to endorse timber sales done under the rubric of promoting forest health. As a result, those organizations inadvertently helped to legitimate the timber industry's framing of the issue. Moreover, they were in a particularly difficult position to address funding for ecological restoration work. Under the Forest Service's timber sale program, a timber company would pay money to the Forest Service for the right to log a timber sale and the Forest Service could then use some of that money for restoration projects. Within that framework, logging was needed to provide the Forest Service with funding to do restoration. However, while the Forest Service received money from timber companies for its timber sales, the total costs of the Forest Service timber sales program exceeded the amount of money that the agency received from the timber companies by about $1 billion per year.[85] Thus, if the timber sale program was abolished, part of the billion dollars saved annually could instead be redirected to create a substantial funding source for ecological restoration work. The National Forest Protection and Restoration Act would provide a reliable source for that funding. However, most national environmental organizations were not willing to endorse the act, so they remained tied to a system under which they would have to accept some logging of national forests as a means of raising funding for national forest restoration work.

The forest health logging frame gained further momentum when it was promoted by the George W. Bush administration. That administration was tenacious in reframing the forest protection debate to argue that restrictions on national forest logging by appeals and litigation were responsible for wildfires, despite a lack of evidence to support this claim.[86] As part of the rhetorical strategy under Bush, the administration's program to reduce public oversight and increase logging was named the "Healthy Forests Initiative."

Bush's initiative got a boost from some unexpected sources—Senate Majority Leader Tom Daschle (D-SD), in conjunction with The Wilderness Society and the Sierra Club. These two organizations were involved in negotiations to allow some public lands logging on a wildlife preserve and a roadless area in the Black Hills National Forest of South Dakota after logging projects there had been stopped by lawsuits by local environmental activists. However, two of the individual plaintiffs to the original lawsuits—one of whom was a Sierra Club member—refused to go along with the national organizations' logging deal. Amid the 2002 summer fire season, Senator Daschle then introduced a rider that overrode the injunctions and imposed the logging plan by exempting the projects in the deal from the National Environmental Policy Act. The rider was purposely timed to boost the reelection campaign of South Dakota's other Democratic senator, Tim Johnson. The Daschle rider became law and Johnson ultimately won reelection. In response, zero-cut activist Denise Boggs declared, "This is yet one more shameful example of the Sierra Club and Wilderness Society looking out for their political connections instead of for the land and wildlife that must actually live with the repercussions of their deal-cutting."[87]

Republicans were quick to point to the precedent set by Daschle's rider. Bush used the rider to help legitimate his Healthy Forests Initiative, declaring "My attitude is, if it's good enough for that part of South Dakota, it's good enough for Oregon."[88] Congress responded as well. Rep. Scott McInnis (R-CO) commented, "After hearing all the hand-wringing from environmentalists downplaying the impact of appeals and litigation, it's nice to see that the highest-ranking Democrat in the nation agrees that these frivolous challenges have totally crippled forest managers. It will be interesting indeed to find out if what's good for Mr. Daschle's goose is also good for the West's gander. We intend to find out."[89] In 2003, McInnis sponsored the Healthy Forests Restoration Act to implement the administration's initiative. While the Republicans had gained a majority in the Senate in 2003, Democrats in the Senate potentially could have used procedural motions to block the act from moving through that chamber. However, Senators Dianne Feinstein (D-CA) and Ron Wyden (D-OR)— who had both been elected with strong support from the Sierra Club— instead chose to broker a legislative compromise that legitimized the Healthy Forests Restoration Act and prevented a Democratic filibuster. Their actions resulted in less environmentally harmful legislation than had originally been proposed, but at the same time it ensured that the act became law.[90] While the Rider from Hell and the Salvage Rider had temporarily shielded some logging projects from litigation, the Healthy Forests

Restoration Act reduced public oversight on national forest management for the long term. In particular, the act limited environmentalists' recourse to the use of appeals and litigation to challenge commercial logging projects done under the pretext of reducing fire.

As they were facing the political challenges presented by the George W. Bush administration, zero-cut groups were simultaneously hindered by unfavorable economic conditions. The burst of the stock market bubble in 2001 hit the grassroots forest protection groups particularly hard because of their reliance on foundation support. After taking heavy losses, foundations and large donors who were heavily invested in the stock market had to scale back their funding. As funding became more limited, the grassroots groups on the margins were most likely to be hardest hit. At the same time, zero-cut activists lost three principal allies in the foundation world. As executive director of the Turner Foundation, Peter Bahouth had been zero cut's biggest funder. However, after he left that position in 2001, the foundation soon turned away from supporting zero-cut groups. Meanwhile, the Foundation for Deep Ecology cut its grant giving to U.S. groups in order to focus on environmental protection in South America. The Levinson Foundation also went through a shift in funding areas and moved away from forest issues. (Patagonia's grantmaking program continued to fund zero-cut groups and was an important lifeline for many of those groups during the early 2000s.)

These economic conditions hit the zero-cut groups particularly hard. A few groups ceased to exist. Heartwood had to close its central office, and other groups significantly reduced their staff size to adapt to these difficult economic conditions. One adaptation by the National Forest Protection Alliance was to collaborate with the better-funded national environmental groups through a new foundation-sponsored coalition called the United Forest Defense Campaign. This coalition was funded through The Wilderness Society and included other national environmental groups. Nonetheless, the coalition did not exclude zero-cut groups, and it offered a program of minigrants of less than $3,000 to grassroots groups. As a result, there were more harmonious relations between zero-cut groups and national environmental organizations in the early 2000s than there had been in the 1990s. However, this harmony came at a price for the zero-cut campaign as this collaboration took place within a framework that fundamentally was not oriented toward ending the Forest Service timber sales program. Instead, it was a defensive campaign intent on fending off the Bush administration's rollback of forest protections. Within this defensive framework, the differences between the moderate national organizations and the zero-cut groups were muted, and in this context it was harder to promote a comprehensive zero-cut alternative to the Healthy Forests Initiative.

Despite these difficulties, zero-cut groups continued to have success in halting individual timber sales on national forests through appeals and litigation during the Bush years. The John Muir Project adapted to the unfavorable economic conditions by shifting to a focus on challenging timber sales in the Sierra Nevada national forests of California. While other organizations had previously monitored these forests, they were not zero-cut advocates and were reluctant to file suit against most timber sales. The John Muir Project had no such reservations and proceeded to win big litigation victories against the Forest Service. Following the John Muir Project's arrival, logging levels on the Sierra Nevada national forests dropped by more than 40 percent between 2000 and 2004.[91] Likewise, nationally, logging levels remained low as grassroots groups continued to appeal and litigate timber sales throughout the national forest system. The switch from the Clinton administration to the Bush administration did not lead to an increase in logging. In fact, overall logging on national forests continued to decline through 2002. That year, logging levels reached their lowest point since the 1930s (see figure 4.1). The next two years brought a slight upturn, but throughout both of Bush's terms, logging levels consistently remained lower than they had been at any point during the Clinton years. Grassroots forest protection activists had a profound impact on the Forest Service's timber sale program that the Bush administration could not undo.

Achieving More by Being Bold

This chapter has examined the history and impact of the segment of forest protection activists working to end all commercial logging on national forests. The zero-cut campaign had two main facets—a legislative component spearheaded by the John Muir Sierrans, and a litigation component carried out largely by new grassroots groups. These two aspects complemented each other, and there was significant overlap between them in terms of the people involved. Together they had a big impact on lowering national forest logging levels. But what is perhaps most striking about the experiences of the zero-cut campaign was the obstructive role played by the national environmental organizations.

The goal of ending all national forest logging was seen as politically unrealistic by national environmental organizations that relied on an insider strategy. Early zero-cut proponents were told by the staff of the nationals that legislation to end logging on national forests would not pass the laugh test, although, as zero-cut advocate Charlotte Talberth observed, "The 'laugh test' is whatever the politicians and the funders backing them are comfortable with." For that reason, advocacy of zero cut was seen as dangerous by groups using an insider strategy that relied on access to those politicians. For example, former Sierra Club Executive Director Michael Fischer feared that if the Club adopted that position, "the club lobbyists would lose the access [the Sierra Club] has in Washington. Committee staffers would simply refuse to meet with them because their own credibility would be affected if they did."[92]

The nationals not only did not endorse zero cut, but at times they actively hindered grassroots forest protection efforts that took a stronger position than they did. As a result, zero-cut activists had to take on the national environmental organizations while they also challenged the Forest Service's logging program. It was only through the extensive efforts of the John Muir Sierrans that the Sierra Club changed its position at the insistence of its membership and over the opposition of many of its leaders. Even after the membership had voted, the Club was still restrained in its advocacy for zero cut. The resources for that campaign were allocated only after difficult struggles by John Muir Sierrans on the board of directors. Moreover, the actions that most significantly moved the zero-cut legislation forward—finding a champion for the bill and a Republican cochampion, doing the economic research to make the case for the bill, and much of the district–level organizing to get congressional cosponsors for the bill—were accomplished by grassroots activists rather than Club staff. And even when the National Forest Protection and Restoration Act had attracted more than a hundred congressional cosponsors, the other national environmental organizations, which had no mechanism to let their members vote for a zero-cut policy, still would not endorse it.

Not only did the National Forest Protection and Restoration Act make far more progress than the insiders had predicted, but their warnings that introducing the bill would harm forest protection advocacy were similarly erroneous. As Chad Hanson recalled,

The major criticism against the end commercial logging campaign before it became a Sierra Club position was, "If the Sierra Club takes that position, all hell is going to break loose. The Sierra Club and the other national groups will lose credibility. We'll start losing on other issues as well. We'll lose funding. We will start getting beat in Congress because we'll be lambasted, et cetera, et cetera. It's going to have these dire consequences. And it will also dramatically hurt our incremental efforts." Everyone has seen and acknowledged since then that this criticism was dead wrong. In fact, not only was it wrong, but exactly the opposite was true. Everything shifted positively in our direction. More was possible. Bigger things were possible. More things could be talked about.

While the national groups other than the Sierra Club did not endorse zero cut, soon after the zero-cut legislation was introduced, those organizations, as well as the Forest Service chief, adopted notably stronger positions on national forest protection, including calls for roadless area protection and an end to all old-growth logging—positions that many of these groups had considered too extreme to support just a few years earlier.

The work of the grassroots zero-cut activists thus helped to strengthen the forest protection advocacy of the nationals, but the grassroots groups did not receive a similar benefit from the work of the nationals. In a complementary relationship between the grassroots and the nationals, the nationals could have helped translate grassroots forest protection goals into legislation. Instead, with regard to the zero-cut campaign, the nationals functioned primarily as gatekeepers against stronger forest protection legislation. Once the Sierra Club was removed as a barrier, the zero-cut legislation made significant progress, due largely to the grassroots mobilization efforts by activists outside of Washington, DC. The grassroots groups discovered that they did not necessarily need the nationals to advance legislation; they just needed the nationals to stop obstructing their efforts. And while the National Forest Protection and Restoration Act had not become law as of this writing, the national environmental organizations did not succeed in passing any other comprehensive forest protection legislation during the period covered by this study either.

Ultimately, the effectiveness of the zero-cut campaign cannot be evaluated solely in terms of legislative outcomes. Grassroots forest protection activists had the biggest impact on national forest logging through their use of the timber sale comment, appeal, and litigation procedures established through the National Forest Management Act and the National Environmental Policy Act. In this context, the nationals again played a problematic role. Despite the much greater resources of the national organizations, the bulk of the work of monitoring and challenging logging on national forests was left to grassroots groups.[93] For example, for the timber sale appeals filed between 1997 and 2003, there was only one large national organization—the Sierra Club—listed among the top seventeen appellant groups, and it was fifth on that list, outranked by four small grassroots groups.[94]

Moreover, the national organizations were involved in political compromises that hindered grassroots activists' ability to do timber sale litigation. As Bryan Bird—who had experience working both for the grassroots group Forest Guardians and the Sierra Club—described:

> Some of these larger groups are so politically sensitive that they often will make decisions that are not in the best interest of the people in the field. They may decide that they are going to let some legislation go forward that makes it much more difficult for us to stop timber sales because it is politically pragmatic at the time. Obviously that's where we get into a lot of conflict.

For example, the Sierra Club staff helped write Senator Hatfield's Rider from Hell in 1989, which overturned the first logging injunction for the northern spotted owl, and The Wilderness Society endorsed it. The Wilderness Society and Sierra Club were also involved with the process that led to the Senator Daschle's rider in 2002 exempting the logging projects in the Black Hills from environmental review and subsequently setting the stage for the Healthy Forests Restoration Act in 2003. Thus the political deal-making by the nationals at times threatened to undermine the legal tools used by the grassroots forest protection activists.

Despite such impediments, appeals and litigation ultimately had a profound effect on logging levels on national forests. From 1988 and 2004, the amount of trees cut plummeted from more than 12 billion board feet down to 2 billion board feet, the lowest level since 1940 (see figure 4.1). During that time, national forests in every region of the country experienced injunctions that halted or dramatically curtailed logging. New grassroots groups such as the Forest Conservation Council, Forest Guardians, Heartwood, John Muir Project, Native Forest Network, Wild Alabama, and many others that emerged in the late 1980s and 1990s were the primary drivers of those appeals and lawsuits. Unconstrained by an insider approach, they were willing to enforce existing environmental laws even when their lawsuits were considered too politically controversial by the national organizations. As a result, grassroots groups made remarkable progress toward ending logging on national forests even without enacting zero-cut legislation. Reflecting on the experiences of the zero-cut campaign, Chad Hanson concluded,

> I think it vindicates those who have said that being bold, being aggressive, and fighting for what you believe in can change political realities and make more things possible. It will not harm protection efforts. It will promote protection efforts. You will not get less. You will get more in every way by demanding more.

5

BECOMING A NATIONAL: THE CENTER FOR BIOLOGICAL DIVERSITY AND ENDANGERED SPECIES LITIGATION

The previous chapter chronicled how forest protection groups used laws such as the National Forest Management Act and the National Environmental Policy Act to dramatically reduce logging on national forests. The most notable early success of this approach was the 1991 injunction shutting down logging on public lands in the Pacific Northwest in order to protect the northern spotted owl, a bird that was designated as threatened under the Endangered Species Act (ESA). That injunction was often incorrectly portrayed as being a result of the ESA. In fact, the primary basis for this injunction was the National Forest Management Act. However, the ESA was subsequently used to stop logging of owl habitat on public lands in the Northwest managed by the Bureau of Land Management, a federal agency to which the National Forest Management Act did not apply. Moreover, the spotted owl issue demonstrated that the protection of endangered species could have potential sweeping impacts on environmental policy.

The Endangered Species Act had implications not only for logging, but also for other environmentally damaging activities on public lands, such as overgrazing by livestock and off-road vehicle use. In addition, the protections provided by the ESA extended to private lands, the oceans, and even internationally. The grassroots groups that emerged during the late 1980s and 1990s looked to the ESA as a powerful tool for environmental protection. While some groups remained focused primarily on national forest protection, others explored how the ESA could be applied to other environmental issues. The latter groups often included the word "biodiversity" in their names—for example, Biodiversity Associates, Biodiversity

Legal Foundation, Greater Gila Biodiversity Project, San Diego Biodiversity Project, and Southern Appalachian Biodiversity Project. Together, groups such as these constituted the grassroots endangered species litigation campaign.

The grassroots group most successful in applying the Endangered Species Act to biodiversity protection was the Center for Biological Diversity. By 2004, the Center's actions resulted in 335 species being protected under the ESA, more than any other environmental group.[1] Indeed, the Center was responsible for 70 percent of all the species added to the list of animals and plants protected by the ESA between 1992 and 2002.[2] It was also responsible for the designation of 43 million acres of critical habitat for those species, an area greater than all of the national parks in the lower forty-eight states.[3] Moreover, the Center's litigation to enforce the protection of species through the ESA resulted in a dramatic reduction in logging on national forests throughout the Southwest, removal of cattle from millions of acres of streamside habitat, exclusion of off-road vehicles from fragile desert ecosystems, limitations on urban sprawl, and restrictions on destructive fishing practices in the oceans.

The Center was also one of the most successful grassroots groups in terms of its organizational growth. In the early 1990s, it was described as "a handful of hippies operating off of unemployment checks."[4] By 2004, the Center had a staff of more than thirty people, offices in six states, and a multimillion dollar budget. It was by then on the cusp of becoming a truly national organization in terms of its scope and influence. In this chapter, I explore the development of the grassroots endangered species litigation campaign, with a focus on the Center for Biological Diversity. I will examine how the Center was able to grow from such modest beginnings, and I will consider the implications of that growth. But in order to understand the achievements of the Center and the grassroots endangered species litigation campaign, I begin by introducing the law that enabled them to accomplish much of what they did—the Endangered Species Act.

The Endangered Species Act

The Endangered Species Act became law in 1973. (For a legislative history of the ESA, see the appendix.) Policy analyst Steven Yaffee described it as the peak of the wave of environmental legislation during the early 1970s.[5] Although it later became quite controversial, at the time the ESA passed unanimously in the Senate, received only four opposing votes in the House, and was signed into law by President Nixon, who declared, "Nothing is more priceless and more worthy of preservation than the rich array of animal life with which our country has been blessed."[6] It grew to become what former Secretary of the Interior Bruce Babbitt called "undeniably the most innovative, wide-reaching, and successful environmental law."[7]

The goal of the Endangered Species Act is to prevent species of wildlife from being driven to extinction by human activities, and to this end the ESA provided potentially powerful tools. Endangered Species Act protection begins once an animal or plant is determined to be at risk of extinction. Those species at greatest risk are designated as "endangered," while those species whose condition is not as severe are designated as "threatened."[8] (There are also "candidate" species that have been found to be imperiled but have not been officially designated as threatened or endangered and thus do not receive the protections of the ESA.) The listing process to designate a species as endangered is established in Section 4 of the ESA. The U.S. Fish and Wildlife Service (FWS), a federal agency in the Department of the Interior, has been given primary responsibility for making these listing determinations and implementing the ESA.[9] The agency is instructed to make listing decisions based solely on scientific criteria and without regard to the potential economic impacts of protecting a species. Under Section 4, once a species is listed under the ESA, the agency must then identify and designate the "critical habitat" areas necessary for the survival and recovery of the species. And the agency develops a "recovery plan" for the species to enable its population to grow to a level where it no longer needs protection and can be "delisted." In total, Section 4 establishes the process to determine which species warrant listing under the ESA, and what sort of protections they will receive. Section 4 is thus the gateway to all subsequent protection under the ESA, and as a result, the listing process has become particularly contentious.

Once a species has been listed as endangered, it is protected by Section 9 of the ESA, which prohibits "take" of endangered species.[10] Take has been defined as "to harass, harm, pursue, hunt, shoot, wound, kill, trap, capture, or collect, or attempt to engage in any such conduct."[11] A court decision subsequently clarified that the definition of "harm" included activities that damage a listed species' habitat in a way that imperils that species. For example, a logging project that sought to cut down the old-growth trees on which spotted owls depend for shelter and nesting would be pro-hibited, even if no owls were directly killed during the logging.

Another part of the ESA that has been central to biodiversity protec-tion efforts is Section 7, which requires any federal agency involved in an activity that might harm an endangered species to "consult" with the FWS. The FWS must then issue a "biological opinion" on the impacts of the activity. The federal agency can only proceed if the project does not jeopardize the survival of a listed species and does not adversely modify critical habitat. Section 7 procedural requirements would prove to be important to many of the legal challenges by grassroots biodiversity groups because this provision has a very broad scope. It covers not only manage-ment of federal lands, but also any project receiving federal money or a permit from a federal agency. For example, if a state highway is being built in part with federal money and impacts a listed species or its habitat, it is subject to the requirements of Section 7. Or if the U.S. Army Corps of Engineers gives a permit for private developers to build on a flood plain, that project is also under the purview of this provision.

The ESA also contains strong citizen-enforcement provisions that have become quite important for its implementation. The U.S. Supreme Court described those provisions as "an authorization of remarkable breadth."[12] Although the FWS is responsible for deciding which species are listed, members of the public can submit "listing petitions" to compel the FWS to consider particular animal or plant species for protection under the ESA. In addition, there is a strong citizen-suit provision under Section 11 that allows members of the public to sue the FWS if it fails to implement the provisions of the ESA. Members of the public can also sue other agencies or private individuals if they are violating the ESA. These suits can result in court-mandated injunctions to halt activities that imperil listed species and court orders to reconsider agency decisions. Moreover, if the mem-bers of the public prevail in this litigation, the defendant (usually the fed-eral government) is then required to pay their attorney fees. As we will see, this provision was particularly important for grassroots groups with scarce resources because it meant that they could find attorneys to work on ESA cases without charge if the attorneys believed that they had a good chance of prevailing in the case and recovering their fees.

Some of the members of Congress who voted for the ESA in 1973 would later claim that they did not realize the full implication of the protections provided by the act, but initially the implementation of the ESA was relatively uncontroversial. The FWS appeared to embrace its new responsibilities.[13] In some cases, environmental organizations such as National Wildlife Federation, Sierra Club, and Environmental Defense Fund applied litigation or the threat of litigation to enforce greater endangered species protection. However, these cases were generally resolved without great controversy, often through negotiated compromises that allowed the development projects at issue to continue with some modification.[14] The main exception was the case of the snail darter.

The Snail Darter Case and the 1978 ESA Amendments

The snail darter case involved a small fish and a small group of activists, but its implications for ESA enforcement were quite large. The snail darter is a three-inch-long perch with no commercial or recreational value. In the early 1970s, its only known habitat was in the one of last remaining undammed stretches of river in eastern Tennessee, but all of that area was scheduled to be flooded by the Tellico dam, a $100 million project that was being built by the a federal development agency, the Tennessee Valley Authority. In 1974, a group of law students and biologists at the University of Tennessee began looking into the condition of the snail darter. After passing the hat among local community members concerned about the impact of the dam on the fish, they raised $29 to fund their campaign to protect it.[15] The next year they petitioned the FWS to have the snail darter protected under the ESA. The agency listed the fish as endangered and designated critical habitat for it, which included the area that would be flooded by the dam. The FWS and the Tennessee Valley Authority engaged in negotiations over the Tellico project, but the Tennessee Valley Authority continued to build the dam.

Fearing that that dam would be completed in the meantime, the snail darter advocates filed suit to stop the dam construction, with University of Tennessee law professor Zygmunt Plater (who was also one of the original snail darter petitioners) volunteering to serve as the attorney for the case. The case ultimately came before the U.S. Supreme Court, which issued a strong ruling in support of ESA enforcement. The Supreme Court ruling affirmed that Section 7 in effect gave the FWS the power to halt actions by other federal agencies that jeopardized endangered species, "whatever the cost."[16] This ruling demonstrated that the ESA protected all species, not just eagles, wolves, and other charismatic species. Moreover, those protections could potentially be enforced even if there were powerful groups that benefited from activities that harmed species. Species listings were required to be based on scientific criteria rather than economic concerns. In these ways, the ruling fit with the growing scientific attention to the importance of biodiversity, and the understanding that the loss of any species could have significant repercussions for the larger ecosystem. At the same time, the snail darter decision also mobilized opponents of the ESA, who grew concerned that that act could harm their economic interests.

In the wake of the Supreme Court ruling, Congress amended the ESA in 1978. Policy analyst Steven Yaffee noted that part of the reason that the 1978 amendments passed was because of divisions that developed among the environmental organizations:

> Up through the spring of 1978, environmentalists were unified and adamant in their opposition to amendments. But in the summer, the unified front fell apart and major groups like NWF [National Wildlife Federation] reluctantly supported an amendment. . . . The NWF had decided that an amendment was likely and the most effective tactic was to contribute to its wording. Other more "radical" groups such as the FFA [Fund for Animals] disagreed and continued to oppose any amendments.[17]

The most notable provision of the 1978 amendments reduced the strength of the ESA by creating an Endangered Species Committee that could exempt federal projects from the act.[18] Because the decisions of this committee could potentially result in the extinction of a species, it was dubbed the "God Squad."

While the 1978 amendments were a bad precedent for endangered species protection, in practice the God Squad did not have a significant impact. For instance, a God Squad was convened to consider an exemption for the Tellico Dam. However, the committee found that the dam was an economic boondoggle and denied the project an exemption. Ultimately, after a few unsuccessful attempts, Senator Howard Baker (R-TN) added a legislative rider to an appropriations bill that exempted the Tellico Dam project from the ESA.[19] The bill was then signed by President Carter (who simultaneously offered apologies to environmentalists for doing so) and the dam was completed.[20] However, not long afterward, the FWS found populations of snail darters outside of the dam area, and the species was eventually down-listed from endangered to threatened status.

The Listing Process and the 1982 ESA Amendments

In the wake of the snail darter case, the FWS became more wary about implementation of the ESA. The FWS's leadership sought to avoid controversial enforcements of the ESA that might upset the members of Congress who held the agency's budgetary purse strings. As part of this strategy, the FWS sought to avoid listing species so as not to have to apply the protections of the act to more species. In 1979, while the agency continued to list some species, it withdrew proposed listing rules for 1,876 species.[21] With the election of Ronald Reagan in 1980 and the appointment of noted antienvironmentalist James Watt as secretary of the interior, this strategy was fully put into practice. In 1981, no new species were listed. While the FWS continued to receive listing petitions from the public, the ESA had not required the agency to respond to the petitions within a specific timeframe, and so the agency was free to ignore these petitions even though it had received at least 4,123 petitions by 1982.[22]

The shutdown of listings and the neglect of citizen petitions provided the impetus for Congress to amend the ESA again. The 1982 amendments set in place clear procedural requirements for the agency to respond to listing and critical habitat petitions within ninety days and to determine whether the listing is warranted within one year.[23] These timelines would later provide the basis for the proliferation of endangered species listings and critical habitat designations in the 1990s. But initially the increase in species listings was more muted, due to a lack of outside pressure. For the remainder of the 1980s, there were very few listing lawsuits filed by environmental groups. Without that outside pressure, there was less incentive for the FWS to list species. And so an audit of the FWS by the Department of the Interior inspector general in 1990 found that there was an extensive backlog of species that were imperiled, but not protected under the ESA:

> We concluded that the U.S. Fish and Wildlife Service had not effectively implemented a domestic endangered species program. . . . Timely progress has not been made toward officially listing and protecting endangered and threatened plant and animal species. Approximately 600 domestic candidate species deemed by the Service to merit immediate protection under the Act have thus far not been officially listed. Also, the Service has identified an additional 3,000

species that are suspected to be threatened or endangered, but action has not been taken to list and protect these plants and animals . . . it may take 38 to 48 years at current listing rates to list just those species now estimated to qualify for protection under the Act. In the meantime, additional species will likely require the Act's protection. . . . We believe that this length of time to list and protect endangered species is not indicative of the "expeditious progress" specified in the Act and could likely result in additional extinction of certain plants and animals during the period.[24]

It was out of these conditions that new grassroots biodiversity protection groups formed to enforce the implementation of endangered species protection in the 1990s.[25]

Jasper Carlton and the Earth First! Biodiversity Project

The grassroots endangered species litigation campaign is largely the legacy of D. C. "Jasper" Carlton. Carlton is an enigmatic figure.[26] He became best known for his role in environmental litigation, but he was not a lawyer. Those who worked with him noted both his fierceness with opponents and his generosity toward those he mentored. He has an uncompromising reputation, but he achieved his biggest successes through negotiation. He rarely sought publicity for his contributions to environmental protection, in large part out of temperament, but also in part out of self-protection. Carlton moved repeatedly between the Southeast and the Rockies, with these relocations "encouraged" by opponents who disliked Carlton's uncompromising approach to activism. On one occasion, a neighbor cut down all of the trees in front of Carlton's home to "celebrate" Earth Day. And on at least two occasions, Carlton was shot at.[27]

By the early 1980s, he was living in Idaho and Montana. It was here that Carlton had his first major success in grassroots endangered species protection.[28] He was working for the Idaho Department of Fish and Game when he learned about the plight of the woodland caribou in that region. Caribou are usually associated with the Arctic tundra of Alaska, but the woodland caribou is a forest-dwelling species once found in the forests of New England, the Great Lakes, and the Northwest. By the 1980s, however, the only remaining caribou in the lower forty-eight states were located in the Selkirk Mountains of Idaho and Washington. Moreover, caribou numbers were well below one hundred individuals, making the woodland caribou one of the most imperiled mammals in the United States, and its population faced further declines due to habitat loss. Woodland caribou depend on the moss found on older trees as their primary food source in winter, but in the early 1980s, the Forest Service was cutting down those forests quickly. In a magazine interview, Carlton recounted his actions in response to the decline of the caribou and reenacted the replies that he received from the federal government:

> We started carrying dead caribou out of the woods. When I carried my third dead caribou out of the woods, the rage set in. I said: "It's over, we're going to federal protection."
>
> "You can't do that," [Carlton] whines, pretending to be a federal bureaucrat.
>
> "Oh yes I can," says Carlton.
>
> I threatened a lawsuit against (interior secretary) James Watt and I won emergency listing of the Selkirk Woodland Caribou (under the Endangered Species Act). We drew up the complaint and on the day we were walking in to file the lawsuit the justice department attorneys met us.
>
> "What do you want?" he asks in a hushed voice, recalling the question of a justice department attorney.
>
> "I want emergency listing and immediate cessation of this timber sale by the Forest Service."
>
> "Mr. Watt will give that to you."[29]

The Selkirk population of the woodland caribou received emergency listing in January 1983, followed by long-term protection under the ESA. That was a particularly remarkable accomplishment given that Secretary of the Interior Watt had recently declared that he would block any new species from being listed.[30]

Carlton's success in protecting the caribou was an inspiration for other grassroots activists. Keith Hammer recalled that prior to becoming a biodiversity activist, he first heard about the caribou listing while working as a logger in Montana. He was particularly struck by the listing because until then he had no idea that there were caribou living in the Northwest: "That inspired me before I had even gone into environmental work." By the late 1980s, Hammer was assisting Carlton with his new Earth First! Biodiversity Project.

The Biodiversity Project grew out of Carlton's involvement with Earth First! in the mid-1980s. Earth First! offered Carlton a community that was supportive of his uncompromising attitude toward biodiversity protection, and the Earth First! Round River Rendezvous provided a space for Carlton to share his litigation experiences with many other activists. Carlton's approach to using environmental laws marked a significant shift in tactics within Earth First!. Hammer said of that approach, "It was a part of [the Earth First!] movement that was about as far away from the monkeywrenching as you could get." On a tactical level, Hammer's assessment is correct. However, in terms of strategy, Carlton's radical approach to litigation shared much in common with Earth First!'s direct action. Indeed, some Earth First!ers called it "paperwrenching" because the appeals and litigation could have such a disruptive effect on environmentally harmful projects.[31] As Kieran Suckling of the Center for Biological Diversity noted, "Basically he was harnessing a lot of very creative, aggressive, gungho energy which otherwise might go into direct actions and put that into the courtroom."

The Earth First! Biodiversity Project itself was little more than a name attached to Carlton's activism. With scant funding, it provided just enough structure to channel Carlton's energies in three ways that would each have a profound effect on biodiversity activism. First, the project offered Carlton a space to further develop and refine his techniques for using environmental law for biodiversity protection. In particular, he worked on a multispecies approach. He did not focus only on large, charismatic mammals but gave equal concern to the fish, amphibians, reptiles, mollusks, insects, and plants that were often overlooked by moderate environmental organizations. He would prepare litigation on a variety of species simultaneously, which would give him increased leverage in negotiating with the regulatory agencies. As Suckling explained, "Jasper is very much about the art of settlement. He really figured out how to overwhelm the agencies, get them over a barrel, and then strike a creative settlement." From this position of strength, negotiation became a tool to extract stronger protections from the agencies rather than a give-and-take process that weakened the final outcome. Suckling noted that some of Carlton's biggest successes came through negotiation.

The second benefit of the Earth First! Biodiversity Project was that it provided Carlton with a forum for training other activists in his strategy and tactics. In many of my interviews, respondents highlighted Carlton's key role as a teacher. As Hammer described,

> He empowered so many people, in terms of helping them learn how to petition for the listings, petition for critical habitat, research the scientific literature on a species (which is necessary for both of those things to occur), and then a paralegal approach to getting all of this stuff together, hiring an attorney, and getting it done. It was incredible how many people I saw Jasper come in contact with through that project. He not only empowered them, he inspired.

Activists taught by Carlton went on to start their own regional biodiversity projects, including the Biodiversity Associates, Center for Native Ecosystems, Florida Biodiversity Project, Greater Gila Biodiversity Project, San Diego Biodiversity Project, Southern Appalachian Biodiversity Project, and the Center for Biological Diversity. As Kieran Suckling observed, "A lot of groups came along right at the same time, or a little bit later, all over the place, and all of a sudden there was this biodiversity movement, which effectively just did not exist before Jasper Carlton in terms of an aggressive activist movement. I really credit him with inventing that entire mode of conservation."

Carlton himself created a somewhat more formal organization called the Biodiversity Legal Foundation in 1991 in partnership with Ned Mudd, an unconventional Alabama lawyer and professional musician who described his musical genre as "redneck electronica."[32] The group would have a significant influence on endangered species policy through the 1990s, though organizationally it remained quite small. Carlton eschewed seeking funding from most environmental foundations, which he saw as a damper on more radical approaches.[33] Instead, the Biodiversity Legal Foundation pushed ahead on limited resources, with Carlton "living like a monk" and relying on pro bono lawyers.[34]

Carlton's third contribution to the grassroots endangered species litigation campaign of the 1990s was his massive multispecies candidate listing case. The suit began with a letter from Carlton to famous consumer advocate Ralph Nader (who had never met Carlton). Carlton complained that the national environmental organizations were not doing enough to protect endangered species and asked for Nader's help. The letter was eventually passed down to Eric Glitzenstein, who was working at the time as a lawyer with Nader's Public Citizen Litigation Group. Glitzenstein agreed to take on Carlton's candidate listing case.[35] Glitzenstein had not worked on endangered species litigation before, but would later contend that this lack of familiarity gave him an outsider perspective that made him more open to the possibilities of Carlton's lawsuit than were the more established environmental lawyers, who dismissed it as unrealistic. Carlton wanted to challenge the Fish and Wildlife Service's candidate species listing backlog. As noted earlier, by 1990, the FWS had hundreds of species in limbo as candidate species qualified for listing as threatened or endangered, but since they were not actually listed, they were receiving no protection. The lawsuit sought to compel the agency to issue listing proposals for all 443 candidate species. In the course of their preparation, Glitzenstein described an unusual opportunity they had to look through the FWS's internal records:

> It had some incredibly useful stuff about the political reasons why they were not listing species and all the data they had about the length of time a species could languish on the candidate list. It was basically a dream for a litigator. Obviously that would never happen again, but it was kind of a demonstration of how far we've all come in some respect, because they had no idea why we wanted this stuff and what we were planning on doing with it.

As the case developed, Carlton and Glitzenstein sought prominent large wildlife protection groups to serve as coplaintiffs on the suit. They were able to gain the support of the Fund for Animals, an animal rights group that had been an early advocate for the ESA. They had a more difficult time gaining endorsements from the national environmental organizations. For example, The Wilderness Society rejected their offer to join the case, though Glitzenstein later discovered that organization taking credit for the case in its promotional materials.[36] Ultimately, one of the smallest of the national groups, Defenders of Wildlife, joined the case as a plaintiff. Meanwhile, in typically publicity-shy fashion, Carlton did not list his name as the lead plaintiff in the case that he had developed.

Glitzenstein filed the suit in 1991.[37] The Bush administration entered into negotiations with the plaintiffs, and by the end of 1992, the case was settled with the FWS agreeing to issue proposed listing rules for all of 443 candidate species by 1996. (The Clinton administration later tried to be exempted from the settlement, but only managed to get the deadline for compliance extended to 1998.)

The multispecies case had two significant implications for grassroots biodiversity protection activism. First of all, it created the conditions for other grassroots groups to see many of these listings to completion. While the FWS issued proposed listing rules for many species under the terms of the settlement, the agency dragged its feet on issuing the final listing rules to complete the listing process. It thus fell to the various new grassroots biodiversity groups to take action to compel the agency to finalize some of the listings. Indeed, about three-quarters of all of the species that were later listed due to legal actions from the Center for Biological Diversity had been part of Jasper Carlton's multispecies case.[38] As a result of that case and subsequent litigation by the Center and other biodiversity groups, more species gained protection under the ESA during the 1990s than at any other point in the history of the act.

Secondly, the victory got the attention of the national environmental organizations. As Glitzenstein recalled,

> It was one of these things that really did shake up the biodiversity world, because my sense is that [the nationals] said, "Wait a second—here we are, these national groups, sitting around. Here's this grass-roots biodiversity activist Jasper Carlton who calls us up and screws with us once in a while, along with these lawyers we never heard of before. And they do this case that basically completely sets different terms of the debate of what could be done or what can't be done."

After that case, the national organizations took the grassroots biodiversity groups more seriously and eventually there were increased opportunities for grassroots activists to get legal assistance from more mainstream environmental attorneys such as the Sierra Club Legal Defense Fund.

Origins of the Center for Biological Diversity

Beyond the many species that he protected, Jasper Carlton's other legacy was the generation of grassroots activists he trained in endangered species litigation. Two of Carlton's most successful protégés were Peter Galvin and Kieran Suckling, who became cofounders of the Center for Biological Diversity. In 1989, Galvin and Suckling met through their involvement in Earth First!. Galvin had been active with Earth First! as a college student in Oregon and then in Arizona. Suckling was a graduate student in philosophy who had gotten involved with Earth First! while traveling in the West. Suckling's dissertation took a backseat when Galvin invited him to join him in working on surveys for the Forest Service, looking for Mexican spotted owls in the national forests of the Southwest. Together with fellow Earth First!er Todd Schulke, they began exploring ways to protect owls and other wildlife that went beyond the direct action tactics of Earth First! As Schulke explained, their "philosophy stayed the same, but the approach changed dramatically." They were particularly impressed with Jasper Carlton's effective use of litigation as a basis for protecting biodiversity.

Carlton conducted a workshop for the young activists that provided the groundwork for what would become the Center's three-step strategy. The first step consisted of an in-depth assessment of the status of imperiled wildlife in the Southwest. This information would provide the basis for extensive litigation as the second step. And the crisis resulting from the litigation would lead to the third step, a fundamental transformation of the behavior of the federal agencies responsible for implementing environmental policies.[39]

To provide an organizational vehicle for this strategy, in 1989 Suckling and Galvin formed the Greater Gila Biodiversity Project, which focused on the Gila River watershed of New Mexico and Arizona. At the time, the national environmental organizations had very little presence in the region, creating an organizational vacuum in which grassroots groups could develop, as Suckling noted, "without bumping shoulders with a lot of the big environmental groups." In this regard, their situation was similar to that of the small groups in northern California at the early stages of the Headwaters campaign (see chapter 3).

Initially, the Great Gila Biodiversity Project's membership consisted only of a few friends. The only foundation grant they got was from the Fund for Wild Nature; so they began the first stage of their project with almost no funding. Nonetheless, during the first two years of preparation as they gathered all available information on biodiversity in the region, they found ways to make do. At first, they had seasonal employment with the Forest Service and unemployment checks during the rest of the year. Housing costs were minimized by living in rural areas in New Mexico in tepees and cabins with no plumbing or utilities (but with a solar-powered fax machine). Suckling would hitchhike to Albuquerque to use the university library. There he would do research in the library all day, sleep in the bushes outside the library at night because he could not afford a motel room, and then return to his work in the library the next morning. Suckling explained, "That was our building-block period, and to this day our success—not just in getting species listed but in winning timber sale appeals, winning lawsuits—has been because we're way ahead of the agency in terms of our knowledge of these species. Once we started doing the litigation, we were ready to go—we had all of the information."[40]

The activists' early benefactor and another cofounder of the Center for Biological Diversity was Robin Silver. Silver was an emergency room doctor in Phoenix, Arizona, as well as a professional wildlife photographer. Being an emergency room doctor gave Silver enough income to carry the nascent Center through its early years and a flexible enough work schedule to play a significant role in its activism. Prior to that time, Silver had been photographing imperiled wildlife for the Arizona Department of Fish and Game. One species of particular interest to him was the Mexican spotted owl, whose populations were falling. In 1989, after learning of a Forest Service plan to log 60 percent of the Mexican spotted owl's habitat, Silver wrote a petition to get that species federally listed under the ESA.[41]

Silver's owl advocacy brought him into contact with Galvin, but at first his main organizational affiliation was the local Audubon chapter. By 1991, Silver was working to protect another imperiled bird, the northern goshawk. However, the Audubon chapter became deterred from pursuing litigation to protect goshawk habitat from logging out of concern that a hostile judge would issue economic sanctions against them. For Silver, this experience demonstrated the limitations of moderate environmental organizations and the need for an alternative. He recalled saying, "We need to have a group that isn't afraid." This goal led him to work closely with Suckling, Galvin, and Schulke in expanding the Greater Gila Biodiversity Project to create the Southwest Center for Biological Diversity (later shortened to the Center for Biological Diversity).[42]

At that point, the activists were ready to enter the litigation stage of their strategy. Both the Greater Gila Biodiversity Project and Silver had appealed Forest Service timber sales through the agency's administrative process, but had found their comments were largely ignored. However, they soon discovered, as Suckling explained, that "litigation is a great equalizer" for grassroots groups: "It's really the great equalizer because at that point you're not simply saying, 'Listen to my voice please'; you're saying, 'I'm going to harness the voice of a federal judge, and that does absolutely have to be listened to.' And so, if you're smart, with relatively little resources you can bring litigation that actually changes the world."

There still remained the issue of how to get the lawyers necessary for this litigation. None of the Center activists were trained as lawyers, and the Center's meager budget could not cover the costs of hiring an attorney. However, the activists discovered a new nonprofit law clinic called Earthlaw that would litigate for them without charge. Earthlaw was started by Mark Hughes, a former Sierra Club Legal Defense Fund lawyer who had become dissatisfied with that organization. In Earthlaw, the Center found a kindred spirit, right down to its frugal use of resources. As Robin Silver noted, Mark Hughes and his family lived "like paupers" in the early years of Earthlaw. More important, Earthlaw was willing to pursue the sort of controversial biodiversity protection cases that interested the Center.

The Mexican Spotted Owl Injunction

The Center's most notable early litigation with Earthlaw involved the Mexican spotted owl, one of three subspecies along with the California spotted owl and northern spotted owl. (The latter was at the center of the logging injunctions in the Pacific Northwest. See chapter 4.) Like other spotted owls, Mexican spotted owls live in older forests, a habitat that was disappearing due to extensive logging in the Southwest, and as a consequence, the population of owls was declining.

In response to Silver's petition, the U.S. Fish and Wildlife Service listed the Mexican spotted owl as a threatened species in 1993. The following year, when the Forest Service did not act on requests to consult with the FWS on the effects of its logging program on the owl, the Center for Biological Diversity, along with a coalition of five other local environmental groups, filed suit against the Forest Service under Section 7 of the ESA.[43] They also sued the FWS for not designating critical habitat for the owl, an action that would have direct bearing on the Forest Service because the ESA forbade federal agencies from doing activities that would adversely modify the critical habitat of a listed species.

In June 1994, U.S. District Court Judge Carl Muecke responded to their litigation by ordering the FWS to designate critical habitat for the owl. The plaintiffs and the agency were ordered to negotiate a timetable for determining critical habitat. However, a leaked memo revealed the agency's plan to release a critical habitat proposal to meet the judge's requirement but then withdraw the proposal soon after. Suckling noted, "We got those memos, put them in front of Muecke, and he was royally pissed. And that's when his first real strong orders came down." [44] Judge Muecke ordered the FWS to designate critical habitat by December 1994. However, when December arrived, it was revealed that the FWS had not done the court-ordered economic assessment of the critical habitat proposal. That omission could potentially further delay its finalization. At that point, Muecke threatened to fine the secretary of the interior and ordered the agency leaders to provide him with a daily account of their work to finalize the owl's critical habitat. The FWS tried to get released from the judge's order to complete critical habitat for the owl after Senator Kay Bailey Hutchinson (R-TX) attached a legislative rider to an emergency Pentagon appropriations bill that put a moratorium on new endangered species designations. The FWS argued that the rider precluded the agency from assigning critical habitat for the owl. However, Judge Muecke rejected that argument, questioning whether Congress could overturn an existing court order without violating the separation of powers provision of the Constitution. Ultimately, the FWS designated 4.6 million acres of critical habitat for the Mexican spotted owl in May 1995 amid Hutchinson's moratorium.

Meanwhile, the Forest Service showed similar resistance to legal efforts to get it to consult with the FWS on the effects of its timber sales on the Mexican spotted owl. Finally, the plaintiffs asked the judge to issue an injunction against the Forest Service. Around this time, National Audubon Society vice president Brock Evans wrote the Center a letter urging them to set aside their lawsuit out of concern that it would give fuel to the antienvironmental efforts of the new Republican majority in Congress. The activists declined his advice. [45] And in August 1995, Judge Muecke halted commercial logging on all eleven national forests in Arizona and New Mexico.

Incidentally, the injunction was issued within days of the passage of the Salvage Rider, which temporarily exempted many logging projects from compliance with environmental laws, including the ESA (see chapter 4). However, while in other parts of the country the Forest Service took the opportunity provided by the rider to allow logging of areas that had previously been protected by court injunctions, the Forest Service was deterred from trying to use the rider to pursue logging in the Southwest amid the Mexican spotted owl injunction. Suckling contended that the Forest Service was afraid that Muecke would challenge the constitutionality of the Salvage Rider on similar grounds in his response to Hutchinson's endangered species moratorium. Such a ruling could have threatened the implementation of Salvage Rider logging nationally. Thus the commercial logging remained halted in the Southwest during the Salvage Rider period.

Moderate environmentalists frequently argue against pursuing controversial litigation by contending that the suit will simply be negated by a legislative rider that exempts the affected projects from environmental laws. Yet in this case, the Center's Mexican spotted owl injunction was untouched by two of the most significant antienvironmental riders of the 1990s. The Center also averted a legislative rider that would have specifically exempted the Forest Service from having to consult with the Fish and Wildlife Service over the Mexican spotted owl. Upon getting word that Senator John Kyl (R-AZ) intended to introduce this rider, Center activists quickly organized a protest at the senator's Tucson office. The event attracted favorable media attention for the activists, and Kyl decided to withdraw the rider.

Ultimately, the Southwest logging injunction continued for sixteen months before the Forest Service completed an assessment of the effects of its logging program on the Mexican spotted owl. By then, the logging levels on the Southwest's national forests had dropped to 46 million board feet from its peak of 511 million board feet in 1989.[46]

Through the Mexican spotted owl campaign, the Center for Biological Diversity had developed a distinct approach to biodiversity activism. As Robin Silver summarized, "The use of science in a very aggressive fashion as the basis for litigation—no matter what the political risk—is what made us different." To illustrate this distinction, Suckling compared the Mexican spotted owl case with similar litigation brought by The Wilderness Society over ESA consultation failures for the national forests in Idaho. In that case, the judge also granted an injunction stopping the logging on those forests. However, Suckling described how the plaintiffs in the Idaho case panicked over the resulting political uproar and asked the judge to withdraw their logging injunction. This comparison underscored how the Center's approach to litigation was undeterred by the prospect of controversy.[47]

Another distinct facet of the Center's approach to litigation evident in the Mexican spotted owl case involved their use of negotiation. While the Center developed a reputation for taking a "no compromise" approach, the litigation inevitably led to court-ordered negotiations between the plaintiffs and the government, and those negotiations were crucial to the plaintiffs' success. As Suckling explained,

> At the end of the day, the event that secured the victory was a negotiation. The suit shut everything down. The suit forced the negotiation. But you know, no judge ever said the timber harvest must decline by 84 percent. All the judges ever say is, "Forest Service, your decision was bad; render a new decision." That's all they're empowered to do. In the short term, they can issue an injunction. If you want permanent policy change, you can't get that through the courts. So what happens is that the courts give you the muscle and the courts create the social crisis. Then you've got to negotiate the long-term solution. I think that's one of the areas where the Center has actually been most consistently misunderstood. I think our greatest skill is to be able to negotiate very powerful settlements that we could never ultimately get through the court system.

At first glance, this point might seem counterintuitive. Throughout the 1990s, there were many examples of national environmental organizations being criticized by grassroots activists for negotiating away environmental protections. Suckling explained that "one of the historical and constantly repeated errors of the environmental movement is to think you're going to get incremental change by negotiating from a position of weakness."[48] In contrast, he highlighted that the Center sought to only negotiate from a position of strength. While the nationals tended to be more cautious and selective in their use of litigation, the Center gained leverage by bombarding their opponents with lawsuits. As Suckling described the Center's approach,

> There's been lawsuit after lawsuit after lawsuit, and we've had victory after victory from the court. We've pounded these people. So they are looking at this situation thinking, "OK, this group has sued us fifteen times over timber sales. I'm now in this negotiation. I might be able to get this timber sale accomplished, but this group is going to come right back and sue me again." So you've become such an annoyance to them that they're thinking, "What can I do to get these people to just go away permanently? What can I do to get them to just leave me alone and not come back with five more lawsuits?" And so they tend to look at the bigger picture. They tend to look for longer-term

solutions. They tend to be willing to give up more in order to get a little bit of certainty somewhere in the world. And a lot of national groups don't have that because they just go with one lawsuit.

Achieving that sort of leverage required persistence, and the Center remained actively involved in ongoing litigation on Mexican spotted owl protection well into the next decade. For example, in 2003, the Center won another victory in court challenging the George W. Bush administration's efforts to reduce the amount of critical habitat designated for the owl. The Center's persistence paid off. Even at the end of George W. Bush's first term, logging levels on the Southwest's national forests remained less than 14 percent of what they had been at the end of the 1980s.[49]

Applying the ESA to New Issues

Beyond its negotiation tactics, what made that Center's approach to litigation distinct was the underlying strategy for social change. As Suckling summarized, "Many groups are hampered by fear of upsetting their congressional connections, their funders, the media, et cetera. While we feel the pull of such things, we daily remind ourselves that social change comes with social tension and our job is to create that dynamic tension, regardless of the pressure to back down or compromise."[50] The Mexican spotted owl litigation served as a template for that approach. And the broad scope of the ESA allowed them to apply it to a variety of other issues as they undertook an intensive campaign of petitions and litigation to get imperiled species from a variety of habitats listed under the ESA.

One of the first listing petitions authored by the Center was for the Southwestern willow flycatcher, a songbird that lives in willow and cottonwood trees along rivers in the Southwest. Following a series of lawsuits against the FWS, the flycatcher was listed in 1995. The Center became increasingly effective at getting species listed under the ESA despite resistance from the FWS. In 1997 alone, the Center was responsible for the federal listing of thirty-nine species.[51] While the national environmental organizations generally focused on large charismatic mammals, the Center came to the defense of "nonsexy species" such as the Sonoran tiger salamander, the San Diego fairy shrimp, and plants such as the Huachuca water umbel. This approach reflected the Center activists' biocentric values as well as their scientific understanding of the ecological importance of these less glamorous creatures.

The Center unleashed a barrage of litigation to enforce protection of the species it had gotten listed. These cases moved beyond stopping logging on national forests to challenging a wide range of other activities. For example, the decline of the flycatcher was linked to the loss of its habitat due to cattle grazing and water development projects, two key industries in the Southwest. Cattle are permitted to graze on much of the western federal lands managed by the Forest Service and the Bureau of Land Management. In this arid environment, cattle concentrate around the few rivers and streams. Intensive grazing in these fragile areas can denude the riparian vegetation along rivers because the cows eat cottonwood and willow seedlings before they can grow into the full-sized trees. The Center's lawsuits to protect the flycatcher and other riparian species ultimately led to the removal of cattle from public lands along more than 300 miles of streams.[52]

The Center's successful 1997 listing of the cactus ferruginous pygmy owl in Arizona brought ESA issues to private land in the Southwest, as the Center sought to protect pygmy owl habitat from urban sprawl around Tucson. In that area, much of the building was taking place in desert washes, which required a permit from the U.S. Army Corps of Engineers. This link to a federal agency provided a basis under Section 7 of the ESA to challenge projects in those areas that could harm the habitat of the owl. Litigation by the Center, with coplaintiffs Defenders of Wildlife and Desert Watch, challenged the Army Corps' permits and led to a shutdown on the issuing of new building permits in owl habitat for almost two years.[53] The pygmy owl injunction would prove to be one of the Center's most controversial actions in Arizona when it halted construction of a high school in owl habitat. Despite the controversy, which the school's proponents presented as a choice of owls versus children, the Center did not back down from its advocacy for endangered species. While the school was eventually built, the prospect of further litigation led the county to develop the Sonoran Desert Conservation Plan, a sweeping agreement for the protection of fifty-three desert species, including the owl, and halted development in up to 80 percent of the remaining open space in the owl's habitat.[54] Moreover, as Suckling noted, the Center's willingness to pursue the pygmy owl litigation despite the school controversy sent a clear message to developers that the Center would be willing to take on any project that threatened endangered species, no matter how contentious. As such, it was a powerful deterrent for other development projects.

Another deterrent for opponents was the Center's courtroom success rate. By the end of 1997, the Center had won thirty-seven out of forty-seven cases, with another thirty-one cases pending.[55] That success rate led a *Washington Post* reporter to comment that "The Southwest Center's litigation record would be the envy of many law firms."[56] Some critics downplayed these numbers by suggesting that the Center simply took on easy cases. To this, former Earthlaw attorney Jay Tutchton responded,

> The Center's litigation is not easy. A lot of those cases were not winners at the beginning. They look easy now because the Center won 90 percent of them, but no one was doing that kind of work before the Center did. Those things had been out there for twenty years with no one challenging them.

At the same time, Tutchton cautioned against looking only at the courtroom success rate to evaluate the effectiveness of the Center. Even when the Center lost in court, the outcome could still lead to more protection for wildlife. An example cited by both Tutchton and Suckling was a lawsuit involving Hoover Dam. In that case, the Center tried to get the water levels behind the dam lowered by twenty-two feet in order to prevent the flooding of a 1,100-acre stand of willow trees that provided habitat for the Southwestern willow flycatcher.[57] The suit could potentially have affected the supply for water to Los Angeles. In court, the two pro bono lawyers representing the Center faced forty-five lawyers representing the federal government and water and power interests from seven states.[58] The Center ultimately lost that case. However, its litigation provided impetus for a multispecies conservation plan for protecting wildlife along the Lower Colorado River.

The Hoover Dam case was indicative of the audacious nature of many of the Center's lawsuits, and its willingness to take on litigation that most other environmental organizations would eschew as hopeless or too controversial. It was also illustrative of the Center's flexible approach. As Suckling explained,

> The other interesting thing about that Hoover Dam case is that we didn't ultimately know what we wanted when we filed it, which many people will tell you is the worst case scenario: "You have to have the

media plan. You have to have the political plan. You have to know the outcome." We rarely do that. And how we've gotten into some of these really big cases, including Hoover Dam, is that we see an opening. We don't quite know where that's going to lead, but we just trust it's going to be a good result. We trust that they're going to respond in a way that I cannot predict at this moment, but they will respond in some manner. And when they do respond, we will be able to take that in a direction that we want to go. . . . You can't plan out too much in advance or you're going to miss a lot of opportunities to get something else that you didn't see. Like in the Hoover Dam case, this whole Lower Colorado River multispecies conservation plan was not originally on the table, but out of the fray, that was something they did.

This flexible approach, combined with unwillingness to be deterred by controversy, enabled the Center to take on a broad variety of campaigns.

Early Organizational Growth

At the same time that the Center's roster of campaigns was growing, the organization was expanding as well. This attention on organizational development distinguished the Center from mentor Jasper Carlton's Biodiversity Legal Foundation. As Suckling described,

> We basically wanted the Center to be as large as possible to do as much work as possible. Jasper thought an organization was a necessary evil for him to get work done, whereas we felt the organization was important within itself—that there needed to be an organization devoted to endangered species that could have members who could write letters and go to meetings, that could be perceived as an authority so that there was someone to quote in the newspaper with some authority on an issue.

The organization grew slowly at first. In rural Reserve, New Mexico, Suckling and Peter Galvin faced severe harassment. Their tires were slashed. Their cars were defecated on. There was even a makeshift wanted poster posted in the county courthouse that depicted a gun's crosshairs over pictures of Suckling and Galvin. They then moved to a cabin in the nearby town of Luna. However, after a year, their landlord was pressured by neighbors into evicting them.[59]

In 1993, they moved to the larger town of Silver City. There they took on their first new staff person, David Hogan. Hogan was a young Earth First! activist who had been involved in the Headwaters Forest campaign and then created a one-person endangered species protection organization called the San Diego Biodiversity Project. When Hogan started with the Center, the group's resources were very limited. In his first year, Hogan was paid $100 per month and had to supplement his meager income with food retrieved from grocery store dumpsters.[60]

By the time the Center had moved to Tucson, Arizona, in 1995, it had grown to a staff of nine. Although the Center's funding had also grown, it retained its frugal approach to resources. At that point, staff members were only paid $400 to $1,000 per month, including Suckling as executive director. The Center's Tucson home office at that time also provided room and board for the staff, with Galvin sleeping in a loft above his desk and other staff in a tepee and a converted bus outside. (In 1997, the Center moved to a new building and left behind the communal living arrangement.)

As one journalist later noted, the Center's low pay "guarantees only the most passionate will apply."[61] Many of the staff members came from a background in Earth First! or other grassroots activism. They generally did not have advanced degrees in biology or law. Like the Center's founders, they were unlikely to be hired by the national environmental organizations, but at the Center they quickly developed a reputation for working very long hours, having an uncompromising attitude toward biodiversity protection, and becoming self-trained experts in wildlife biology and environmental law. In addition to writing listing petitions, staff would often do much of the preparation for litigation, which made it easier to find lawyers to take their cases pro bono.

By 1996, the Center had a budget of more than $380,000, with about half of that funding coming from foundations. At first, however, the activists were inexperienced with grant writing, and the responses from most foundations were not enthusiastic. After receiving an early Center proposal, the Harder Foundation officer replied, "That's not a grant proposal; that's a ransom letter."[62] Harder did not fund them that year, but offered guidance in grant writing, and later that foundation did support them. The Foundation for Deep Ecology was one of the Center's earliest supporters. More substantial funding then came from the Turner Foundation under Executive Director Peter Bahouth, a former Greenpeace leader who became a key supporter of the Center. By 1996, the Turner Foundation was giving them $45,000 per year.

However, the largest block of funding came from Pew Charitable Trusts. When the Center initially sought funding from Pew, the foundation staff indicated that they were interested in funding a multigroup forest coalition. This led to the creation of the Southwest Forest Alliance in 1995, with the promise of half a million dollars from Pew over two years.[63] Through contract work for the alliance, the Center obtained a sizable portion of this funding. By 1996, it had received more than $70,000 through the alliance.[64]

The Center had a decidedly more positive experience with Pew funding than the zero-cut forest protection groups (see chapter 4). Indeed, while those groups were criticizing Pew, Suckling published a defense of the foundation in the *Earth First! Journal*, originally titled, "Is It Pew or Is It You?"[65] In this article, Suckling argued that some environmental groups were complicit in the problems attributed to foundations when those groups were willing to change their activities to fit the perceived preferences of a funder. By contrast, Center staff took a more cavalier approach to foundations. For example, Robin Silver said, "We didn't use the Pew money to follow their plan. We did what we thought was necessary to shut down logging. And it was the only successful Pew [forest] campaign in the country."

The Center enjoyed greater latitude with foundations than many other groups. The Center's extensive litigation resulted in direct and tangible increases in environmental protection, which particularly appealed to foundation staff who needed to demonstrate measurable outputs to their board of directors. This situation allowed the Center to get the benefits of foundation funding without much of the moderating effects. Suckling wrote in 1996, "As our membership grows, we want to decrease our relative dependence on foundations, but for the time being, they definitely make our work possible."[66]

Relationship to the Nationals: The Endangered Species Coalition and ESA Reauthorization

The Center developed a multifaceted relationship with the national environmental organizations. Center staff publicly criticized some national organizations when those groups endorsed political compromises that undermined the Center's work. At the same time they also encouraged some national organizations to sign on as coplaintiffs to Center litigation in the hope that the visibility of these groups would bolster the credibility of their lawsuits in the eyes of the judges. Generally though, in its early years, the Center for Biological Diversity's relationship to the national environmental organizations was similar to the "never mind the nationals" approach of the Headwaters groups. As Suckling summarized, "Initially our thought was that political work was what some of the other groups did. We had this idea that we create the policy means, and those other groups with the big budgets—their job is to do the policy work. That's the division of labor, and we left it to them."

However, by the mid-1990s, Suckling and other grassroots biodiversity activists were deeply dissatisfied with the work of the national organizations in Washington, DC regarding the reauthorization of the Endangered Species Act: "They weren't holding up what I believe was an implicit contract that the local and the regional groups do the on-the-ground stuff, and the big groups do the political lifting and represent us in some manner there." This problem stemmed in part from what the grassroots activists saw as a "Beltway mindset" among environmentalists in Washington, DC. From that perspective, environmental victories such as the Center's injunctions were seen primarily in terms of their ramifications in Congress. This mindset focused on fears of political backlash in response to litigation victories. As Suckling described,

> Ultimately, they look at it all as headaches for them. When they pick up the newspaper, and they read, "Injunction shuts down 100,000 acre timber sale," the first thought that comes into their head is, "Oh, what a pain in the ass. I'm going to have to deal with a [legislative] rider [trying to overturn the injunction]." So they actually see every environmental victory as a problem that complicates their life.

Even before the Mexican spotted owl victory, the logging injunction in the Pacific Northwest for the northern spotted owl had brought national attention to endangered species issues. At the same time, opponents had used it as a rallying cry for trying to weaken the Endangered Species Act. In the face of this controversy, the national environmental organizations leading the Endangered Species Coalition (ESC) decided not to push for the scheduled reauthorization of the ESA. Frustrations over the coalition's inaction ultimately led the Center to intervene in the ESC to try to transform the behavior of the nationals, much as the John Muir Sierrans sought to do within the Sierra Club (see chapter 4). However, to understand this intervention, it is necessary to first provide some context on the dynamics of the ESC leading up to that time. Indeed, the troubled history of the coalition offers a microcosm of grassroots concerns about the political strategies of the nationals.

The Endangered Species Coalition was created to coordinate the efforts by the national environmental organizations working on the reauthorization of the Endangered Species Act. Reauthorization meant that the ESA would likely be amended in ways that could strengthen and/or weaken it. As described earlier, the ESA had gone through three previous reauthorizations in 1978, 1982, and 1988, and was scheduled for reauthorization again in 1994. In preparation, the ESC hired its first executive director, Karyn Strickler, in 1993. Strickler did not come from the ranks of the national environmental organizations but instead had previously led grassroots campaigns around women's reproductive rights. As executive director of ESC, she began applying that grassroots organizing approach to the Endangered Species Act's reauthorization.

When Strickler was hired, Clinton had just become the first Democratic president in twelve years and the Democrats controlled both houses of Congress. As such, she saw these political conditions as a unique opportunity to reauthorize the ESA and warned that conditions would likely become less favorable after the next election cycle. To capitalize on the opportunity, she emphasized building a network of local activists to do district-level organizing and brought more grassroots environmental groups into the coalition. However, these grassroots groups had no direct role in the coalition's decision making. Instead, decisions were made by a steering committee consisting of representatives from ten of the largest environmental organizations.[67]

Strickler would later characterize the approach of the steering committee as an example of what she called the "Do Nothing Strategy." In her essay on this topic, she characterized the organizations composing the committee as "giant bureaucracies where self-perpetuation, the quest for funding from large foundations, and the desire for a seat at the political table has replaced environmental protection as the primary goal."[68] The Do Nothing Strategy manifested itself in the ESC as an unwillingness by the steering committee to push for legislation to reauthorize the Endangered Species Act. Strickler recalled that steering committee members indicated they did not want to press for a vote on a bill because they "didn't want to force their friends in the U.S. Congress to vote on a controversial issue in an election year."[69]

By contrast, grassroots biodiversity activists around the country were urging Strickler to get the ESA reauthorized immediately. After the coalition steering committee decided to not pursue a reauthorization vote, Strickler sent a memo to the grassroots activists suggesting that they should convey their thoughts on the issue directly to the steering committee members. In response to that memo, the steering committee immediately fired Strickler. All of the women on the ESC's staff—half of the total staff—resigned in protest over the committee's action. Subsequently, frustrated grassroots activists pushed to have the steering committee meetings opened up to include representatives of grassroots groups. Jasper Carlton of Biodiversity Legal Foundation was the first to be added, and he was later joined by the Center for Biological Diversity.

Meanwhile, in the November 1994 elections, the Democrats lost control of both houses of Congress for the first time in forty years, and the Republican majority mobilized around the "Contract with America," which included anti-ESA provisions. This political shift threatened to undermine the insider strategy of those national environmental groups that had focused on building their access to Democratic politicians. These organizations were afraid that they would be powerless to stop the Republicans from gutting the Endangered Species Act. And, indeed, within months of the beginning of 1995 legislative session, Senator Kay Bailey Hutchinson (R-TX) attached a legislative rider to the defense supplemental appropriations bill that put a temporary moratorium on new endangered species listings, until it was rescinded by President Clinton in mid-1996.[70]

In early 1995, the Endangered Species Coalition hired a new executive director, former congressman and forest protection advocate Jim Jontz (see chapter 4). Like Strickler, Jontz embarked on a strategy based on mobilizing grassroots activists at the district level. However, Jontz's organizing approach quickly ran afoul of one of ESC's key funders, Pew Charitable Trusts. Pew's subsequent intervention in the ESC has often been cited as a vivid example of the problems of foundation influence over environmental advocacy.[71] Pew encouraged the ESC to develop a resource-intensive media campaign using the Pew-funded Environmental Information Center (later National Environmental Trust), led by Phil Clapp, to direct that campaign. Pew put forward $600,000 for the ESC, with other funders adding another $400,000. However, Jontz and Clapp's divergent approaches led to a falling out between them. Pew offered substantial funding to the coalition on the condition that Jontz would leave the ESC and Clapp would replace him.[72] Initially, the ESC voted to retain Jontz, but after Pew pulled its funding and other foundations followed suit, the steering committee reversed that decision. Soon after, citing fiscal concerns, National Audubon Society—which had served as the fiscal sponsor for the ESC—unilaterally decided to fire the entire ESC staff.[73] In September 1995, Defenders of Wildlife stepped in to become fiscal sponsor for the ESC, but by then the ESC had lost most of its funding and Pew's resources were going to Clapp's organization.[74]

Pew's intervention led to organizational disruption within the ESC at a time when it was already facing a hostile Congress. This external threat also fostered serious rifts in the coalition. Ironically, the prospect of anti-ESA legislation from opponents in the Republican congressional leadership led the national environmental organizations to take a more active role in pushing for endangered species legislation than they had when Democrats had controlled Congress. However, with their traditional allies now in the minority, some environmental groups were willing to endorse "compromise" legislation that would weaken endangered species protection in order to get sufficient support from Republicans. The most extreme example of this approach involved secret negotiations between the Environmental Defense Fund (EDF) and a Republican congressman.

At the time, EDF was already viewed with suspicion by other environmental groups for its willingness to collaborate with Republican politicians and corporations known for antienvironmental policies. In the 1980s, EDF had become one of the most visible advocates of business-friendly "third-wave environmentalism," which included market-oriented strategies such as pollution trading. Then in the early 1990s, EDF formed a partnership with McDonald's at a time when McDonald's was being targeted by environmental justice organizations over the corporation's waste output. EDF and McDonald's crafted a plan for the corporation to change some of its packaging, but the changes were far less than what the environmental justice groups had sought. Nonetheless, EDF and McDonald's both put out press releases touting these changes as an environmental victory. In this regard, EDF was seen as using its reputation as an environmental organization to provide "green" legitimation to the actions of environmentally damaging corporations, while at the same time undercutting the work of other environmental groups.[75]

Nonetheless, EDF played a prominent role in the Endangered Species Coalition. EDF's primary staff person on endangered species issues, Michael Bean, was seen as one of the top experts on the ESA by many environmentalists in Washington. During Strickler's tenure at the ESC, Bean was the chairman of the steering committee. Reflecting EDF's business-friendly approach, Bean advocated for an "incentive-based" approach to endangered species protection on private lands, where the main incentive to the landowners was reduced enforcement of the ESA. EDF was also more amenable to negotiating directly with representatives of industries that were seeking to weaken the Endangered Species Act. EDF was the most conservative group within the Endangered Species Coalition, but Bean was able to exert disproportionate influence on the coalition, especially once the Republicans controlled Congress. Steering committee members felt pressure to accommodate Bean for fear that EDF would leave the coalition and then cut a deal with the Republicans on legislation that would weaken the Endangered Species Act. A year later those fears would prove correct, albeit in the reverse order.

In spring 1996, it was discovered that EDF (along with the Center for Marine Conservation, Nature Conservancy, and World Wildlife Fund) had been involved in secret negotiations over crafting endangered species legislation with a Republican congressman and ESA opponents, including logging corporations, water development interests, and the National Association of Realtors.[76] Bean contended that the secrecy was necessary to bring all of these parties to the table to negotiate.[77] However, the secret negotiations violated an agreement among the ESC members to keep each other informed of any efforts to amend the Endangered Species Act. This agreement was intended to prevent a divide-and-conquer strategy whereby ESA opponents co-opted a few of the more conservative ESC members into legitimizing a weaker bill than other coalition members would accept, yet the ESC now faced that exact situation.

The steering committee meeting of the ESC following the revelation of the secret negotiations was the first to include Kieran Suckling of the Center for Biological Diversity. Suckling recalled that during that meeting he was amazed to find that not only was Bean unrepentant over breaking the coalition's agreement, but that the members of the other national groups seemed unwilling to take action against EDF over its violation. At that point, Suckling put forward a motion to expel EDF from the coalition because of its actions. Other committee members initially resisted acting on the motion, but Jasper Carlton seconded it, and eventually Suckling and Carlton were able to compel the committee members to vote on it. At that point, the motion passed almost unanimously, but before the vote was over Bean walked out of the meeting, ending EDF's participation in the coalition.[78]

Suckling's successful motion to expel the Environmental Defense Fund was a watershed moment for the Endangered Species Coalition and marked a notable change in direction. Bill Snape, the Defenders of Wildlife representative on the ESC at the time, later commented that the expulsion of EDF "was better for the coalition because it allowed the coalition not to be so focused on being so defensive and not being so confused in its messages about what it wanted."[79] The ESC switched gears to develop and promote a proactive endangered species protection bill called the Endangered Natural Heritage Act, which was endorsed by grassroots groups such as the Center for Biological Diversity. With the ESC hindered by limited resources to promote it, the bill stood little chance of becoming law. Nonetheless, this legislation represented a significant shift by the ESC toward asking for what it actually wanted rather than following the Do Nothing Strategy or supporting compromise bills that would weaken the ESA. The coalition members also organized to fend off a Republican bill that was endorsed by EDF even though it was widely viewed as undermining endangered species protection. With the defeat of that bill (along with the setbacks for Republicans in the 1996 elections), the immediate legislative threats to the ESA diminished for the remainder of the Clinton administration.

Although the ESA had not been formally reauthorized, Congress would continue to fund the act each year without amending it. The ESA had not been dismantled by its opponents nor was it weakened by compromise legislation that had been endorsed by some national environmental organizations. A legislative stalemate maintained the legal status quo. And with that status quo, grassroots groups still had unfettered access to the legal tools that they had been applying effectively to enforce endangered species protection.

Relationship to the Clinton Administration: Struggles over Listings and Critical Habitat

With the legislative efforts of anti-ESA Republicans neutralized, the most tangible threats to endangered species protection came from policies of the Clinton administration—particularly Secretary of the Interior Bruce Babbitt—to curtail ESA implementation. Babbitt pursued this approach ostensibly to assuage opponents of endangered species protection and thereby dissuade them from undermining the ESA. While many moderate environmentalists lauded Babbitt, grassroots biodiversity activists were outspoken critics of his policies. Kieran Suckling's ironic description of Babbitt's strategy was that "he was going to save the ESA by gutting it."

Some of Babbitt's policies made it more difficult to protect endangered species over the long term. One of the defining features of the endangered species program during the Babbitt years was its use of habitat conservation plans (HCPs). Under an HCP, the recipient can receive a permit to harm or kill (i.e., "take") endangered species in exchange for some mitigation measures, though in practice the mitigations generally did not compensate sufficiently for the harm to the species.[80] The HCP program was created by the 1982 amendments to the Endangered Specis Act, but was not actively promoted until the Clinton administration. Only fourteen HCPs had been issued by the time Clinton took office a decade later.[81] However, by 2000, at the end of the Clinton terms, hundreds of HCPs had been issued covering almost 12 million acres of land. Many of the HCPs were given to large corporations, such as International Paper, Wal-Mart, Enron, and Shell Oil. These HCPs often covered thousands of acres, and some were issued for eighty to one hundred years.[82] Moreover, in 1994, Babbitt developed the "No Surprises" policy for HCPs, which ensured that landowners would not need to do any additional mitigation measures, even if the species continued to decline or if other species on their property became endangered. This policy had the potential to preclude new ESA enforcement measures on millions of acres for up to a century. It was challenged primarily by a tiny grassroots biodiversity protection group called Spirit of the Sage Council led by Leona Klippstein, with Eric Glitzenstein serving as the pro bono attorney.[83]

Babbitt was also responsible for policy changes that made it more difficult for environmentalists to get species listed. Most notably, the Department of the Interior issued a policy that prohibited citizen-listing petitions for species that were already on the candidate list.[84] The Fish and Wildlfie Service did not have to provide any protections for a candidate species and was under no clear timeline to officially list a candidate species as endangered, so candidate status became a limbo for species listings. The agency could grant a species candidate status as a way to shield it from citizen petitions that could otherwise open the door to litigation and a court-mandated listing. This tactic was evident in the agency's decision to give candidate status to the Gunnison sage grouse, a ground-nesting bird whose listing would likely have widespread implications for livestock grazing, oil drilling, and off-road vehicle use in Colorado. As the FWS field supervisor for Colorado explained in a memo:

> If petitioned, the Service would have to consider listing the sage grouse and within a year would make a finding on a listing decision. . . . [T]he Service decided that placing the Gunnison sage grouse on candidate list now is the best option. Placement of the Gunnison sage grouse on the candidate list now will likely allow for continuation of the working groups as they are currently operating for several years. In contrast, if we receive a petition, legal charges may result in a threatened or endangered listing in the near future.[85]

In the years that followed, the Center for Biological Diversity and other groups would work to build the case law for a sweeping lawsuit filed in 2005 challenging the FWS's candidate policy for the 283 species in that category at that time. If successful, this suit could unleash a flood of new listings akin to the effect of Jasper Carlton's multispecies lawsuit in 1991. However, in the meantime, the FWS's policies were at least partially effective in curtailing listings. The Center continued to get species protected through petition and litigation, but the overall rate of listings declined continuously from 1997 though 2003 (see figure 5.1).

This lower listing rate also reflected the Center's shift in emphasis to getting critical habitat designated for listed species. Critical habitat, like species listing, was quite controversial with ESA opponents, and the FWS was reluctant to designate it. From 1992 to 1999, Babbitt approved critical habitat for only six species, and these were cases in which he was under court order to do so. By comparison, during the Reagan administration, fifty species received critical habitat, and that was done without court orders.[86] The hurdles that the Center faced to get critical habitat designated were another example of the ways that the FWS during the Clinton era hindered endangered species protection.

Under the provisions of the Endangered Species Act, the FWS is supposed to designate critical habitat at the same time that it lists a species. If the agency does not feel it has sufficient information, it can delay determining critical habitat for up to a year. The only other exemption is that the FWS can elect not to designate critical habitat if it is "not prudent" to do so because the designation will either harm or not benefit the species.

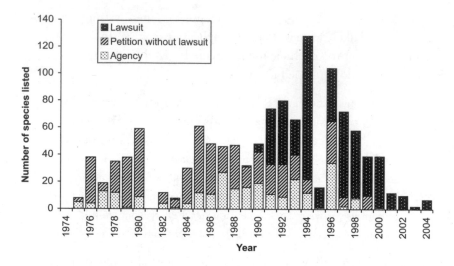

Figure 5.1
Number of species listed under the ESA by year due to agency action, petitions, and lawsuits, 1974–2004.

This exemption was created to protect endangered plant species when mapping the location of the remaining plants as part of the critical habitat designation would leave them at risk of being found by rare plant collectors. However, the FWS applied this exemption to a wide range of animal and plant species that were not at risk from collectors. The "not prudent" exemption provided a way for the FWS to avoid controversial critical habitat designations much as the candidate category provided a loophole for controversial listings. During the second half of the Clinton administration, the FWS consistently used the exemption to deny critical habitat designations, starting with the red-legged frog (the first listing following the Senator Hutchinson's ESA moratorium) and continuing for almost all of the next hundred listed species.[87] There were various legal challenges against the application of this exemption to various species, leading to a lawsuit by the Center for Biological Diversity—done in conjunction with Christians Caring for Creation (see chapter 4)—which challenged the exemption policy as a whole. The success of this case opened the door for overturning the previous denials, and designations of critical habitat consequently skyrocketed (see figure 5.2). Critical habitat lawsuits can only be filed within six years of the species being listed, so this type of case became a major focus of activity for the Center as the deadline approached for the many species that had been listed in the early to mid-1990s. At the beginning of 2000, only 10 percent of all listed species had received critical habitat. But due to litigation, over the next five years, that figure grew to 37 percent.[88] By 2004, the Center was responsible for the designation of more than 43 million acres of critical habitat for endangered species.[89]

Despite these victories, Babbitt and the FWS director Jamie Rappaport Clark were outspoken critics of critical habitat. (Clark continued to critique critical habitat after she left the FWS and became a vice president for the National Wildlife Federation and then Defenders of Wildlife in the early 2000s.[90]) They contended that critical habitat did not provide significant additional protections for endangered species and that the costs of designating this habitat were draining money away from other endangered species work. However, both of these concerns were based on static notions of political reality. Rather than abandon critical habitat, the Center worked on changing the political realities.

The first of these claims—that critical habitat did not provide significant additional protection—stemmed from a questionable rule modification by the FWS during the Reagan administration. Under the ESA,

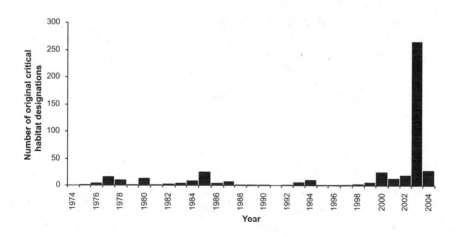

Figure 5.2
Number of original critical habitat designations under the ESA by year, 1974–2004.

federal agencies cannot cause "adverse modification" of critical habitat. This provision precluded not only activities that directly harm the species, but also activities that hurt the recovery of the species. However, in 1986, the FWS made a minor word change to its regulation that had the effect of undercutting these protections. The agency inserted the word "both" into the definition of "adverse modification" so that it now covered activities that harmed "*both* the survival and recovery of a listed species."[91] The addition of the word was nonsensical because an activity that harmed the survival of a species inevitably also harmed that species' recovery, whereas an activity that harmed the recovery of the species did not necessarily threaten its immediate survival. This change in wording had the effect of eliminating the role of critical habitat in species recovery because critical habitat would now only protect areas from activities that immediately jeopardized a species. It would not preclude activities that solely harmed a species' recovery. Since there were already other provisions of the ESA that forbade actions that directly jeopardize a species, the FWS argued that the benefits provided by critical habitat were negligible and thus contended that designation of critical habitat should not be a priority for the agency. This led to a near shutdown of new critical habitat designations in the late 1980s (see figure 5.2). Once he took office, Secretary Babbitt did not correct the FWS's alteration of its definition of adverse modification, and the agency continued to avoid designating new critical habitat until litigation made this policy unfeasible. In contrast to Babbitt's approach, the grassroots biodiversity groups worked to undo the detrimental redefinition of critical habitat. By the early 2000s, the Center and a grassroots group called the Gifford Pinchot Task Force had both won court cases overturning the 1986 language and reaffirming the benefits of critical habitat through its role in species recovery.

A similar pattern was evident with regard to the claim of critical habitat designation being an economic burden for the FWS. While Babbitt and Clark sought to limit critical habitat designations to match existing funding levels, the Center criticized the Department of the Interior's budget requests and called for larger appropriations for the endangered species program to better match the actual needs of that program. The FWS did not receive sufficient funding to designate critical habitat for the species that had already been listed, let alone to address the hundreds of candidate species waiting to be listed and receive critical habitat. At first glance, the responsibility for this insufficient appropriation might appear to lie with Congress. However, the Center's research found that Congress had generally given the FWS almost all of the funding it requested for endangered species listings, and sometimes even more. Moreover, the FWS had dramatically increased its funding for landowner compensation projects, while its requests for listing funding had remained virtually static.[92] In other words, the FWS appeared to be using limited appropriations as a means to tie its own hands when it came to listing and critical habitat.

Furthermore, Babbitt intervened to prevent outside pressures from expanding that budget. In 1997, Babbitt made an administrative rule change that precluded the agency from spending more on listings and critical habitat than Congress had appropriated for that line item. Thus when faced with a surge of demands for new listings and critical habitat designations, the agency could not draw on other parts of its budget to fulfill these needs. Had Babbitt not made this rule change, the growing demands from litigation could have produced a fiscal crisis within the agency that could have compelled it to request a budget increase to address the backlog. However, Babbitt's rule change insulated the rest of the agency from this pressure and thus defused the impetus for change. The agency could instead simply limit itself to appropriating enough money to cover the court-ordered listings and critical habitat, and not do anything else. Babbitt's actions thus enabled the agency to do the absolute minimum level of activity for the protection of new endangered species.

As noted earlier, some environmentalists saw Babbitt's efforts to limit implementation of the ESA as a strategy to assuage its critics and thus protect the ESA from being dismantled. However, the extent of Babbitt's opposition to critical habitat seemed to go well beyond what might be necessary under this strategy. For example, Babbitt asked Congress to cut the amount of funding it planned to give for critical habitat. He also actively encouraged opponents of the ESA to introduce legislation to abolish critical habitat. And when an alternate piece of legislation was proposed that would have provided funding to the FWS to resolve the critical habitat backlog, Babbitt objected to that legislation and it failed.[93]

Likewise, Fish and Wildlife Service Director Jamie Rappaport Clark's actions at times seemed more oriented toward weakening ESA implementation than protecting it from opponents in Congress. In 2000, Clark used an exhausted listing budget as the rationale to halt new listings, even though she and Babbitt had sought to have the listing budget lowered in the first place. Clark's action was particularly ironic because during Clinton's eight years in office, the only significant legislative setback for the ESA was Senator Hutchinson's temporary moratorium on endangered species listings in 1995. Yet at the end of the Clinton administration's second term, a member of that administration, Jamie Rapport Clark, voluntarily created a similar moratorium when no legislation compelled her to do so.

Regardless of Babbitt and Clark's intentions, their actions illustrated that the Clinton administration was at best an unreliable ally for biodiversity protection advocates. While the election of George W. Bush would bring new difficulties, the policies of Babbitt and Clark represented some of the main impediments for biodiversity activists in the 1990s. Those impediments were primarily overcome by litigation from grassroots groups.

Expanding beyond the Southwest

At the same time that the Center for Biological Diversity was putting more energy into critical habitat litigation, the late 1990s also brought significant organizational developments within the Center. For one, it moved beyond being simply a regional group. It began opening offices in California and other locations outside of the Southwest and removed "Southwest" from the organization's name to reflect its broadened scope.

The Center also started hiring its own lawyers. At first glance, this move might seem surprising. Staff attorneys are generally associated with large environmental organizations that can afford to pay high salaries. By comparison, the Center had been able to have a big impact with few resources through its use of pro bono lawyers. However, that approach was encountering some limitations. Most notably, one of the Center's main sources of outside lawyers, Earthlaw, had been absorbed by Earthjustice (formerly Sierra Club Legal Defense Fund). Earthjustice would continue Earthlaw's legal assistance to the Center, but it was not clear that it would do so at the same levels. Moreover, the Center's rapidly expanding litigation case load risked outstripping the supply of readily available pro bono environmental lawyers. Therefore the Center began to hire its own staff attorneys, but as with many other of its facets, it took a decidedly unconventional approach.

The first lawyer hired was Brendan Cummings, who did not fit the mold of a typical attorney. His long hair stretching down the length of his back immediately distinguished him from other lawyers, but even more distinctive than his appearance was his background. Like many other Center activists, he had roots in the radical wing of the environmental movement. He had been involved with Sea Shepherd, the ocean-going counterpart of Earth First! (see chapter 1). He had also participated in the Headwaters Forest campaign, and later became part of the legal team for the Headwaters activists who had been pepper-sprayed (see chapter 3). In 1998, fresh out of law school, he intended to start his own biodiversity project, modeled on the Center, to focus on marine issues. However, before he could do so, he was hired by the Center to provide legal counsel on a half-time basis. He would also litigate cases for the Center as an independent pro bono attorney.

Although a graduate of the prestigious Boalt Law School of the University of California, he accepted a starting part-time salary of $10,000 per year. His cramped apartment became the Center's first San Francisco Bay Area "office." With a mattress on the floor in the corner and a low ceiling slanting overhead, it was the urban equivalent of the Center's early makeshift offices, but this minimal structure was sufficient to attract legal interns from Boalt Law School, including Kassie Siegel, a law student who was also active in the Headwaters Forest campaign. Upon graduating, Siegel became the Center's first full-time staff attorney.[94] (Cummings later switched to full time as well.)

Cummings began applying the legal strategies that had worked so effectively in the Southwest to California. For example, the Mexican spotted owl case had been based on the failure of the Forest Service to consult with the FWS on activities that impacted endangered species under the requirements of Section 7 of the ESA. Cummings successfully brought a similar case against all four of the national forests in Southern California. That was followed by a case that applied a Section 7 claim to the desert lands in California managed by the federal Bureau of Land Management (BLM). Kieran Suckling later described the desert case as "possibly our biggest victory." The lawsuit transformed the BLM's management of 11 million acres in an area known as the California Desert Conservation Area. This area was home to twenty-five endangered species, including the desert tortoise, the Peninsular bighorn sheep, and a desert flower called the Pierson's milk-vetch. However, the BLM had not consulted with the Fish and Wildlife Service on the potential effects of its management plan for the California Desert Conservation Area on those species. In March 2000, the Center filed suit against the BLM for its failure to consult under Section 7.

Despite the large area involved, the Center relied on a small team to take on the campaign. In this regard, the desert case was indicative of the decentralized structure of the organization. There were three primary activists involved. Cummings not only served as a lawyer in the case, but was also fluent in the science of the issue and was active in media outreach.[95] Because of the size of the case, Jay Tutchton of Earthlaw/Earthjustice joined Cummings as the other half of the Center's legal team. Daniel Patterson served as the Center's organizer for this campaign. Patterson, an Earth First! activist who had previously worked for the BLM, was well acquainted with the agency's internal dynamics. His role in the campaign included public outreach, which brought in Public Employees for Environmental Responsibility—a support group for agency whistleblowers—and the Sierra Club as coplaintiffs. The small team of Center activists was given autonomy to negotiate the case without extensive oversight from the group's directors. Executive Director Kieran Suckling highlighted the value of this decentralized arrangement in contrast to the top-down approach to litigation management among the national environmental organizations. From Suckling's perspective, the activists who were most directly involved in the issue were in the best position to make decisions in settlement negotiations since they had the greatest knowledge of the situation. They were also directly invested in saving that place and thus were less likely to negotiate away protections unnecessarily. The decentralized approach allowed the Center team to be nimble and opportunistic in their settlement talks with the BLM.

The result was a settlement that transformed the BLM's management of 11 million acres of California desert. This agreement contained a wide variety of provisions, including limitations and removal of livestock on 2 million acres of land where cattle had been consuming the vegetation on which the desert tortoises depended; a ban on new mines in 3.4 million acres of endangered species habitat; and new protections for the Peninsular bighorn sheep. It was the provision of the settlement involving the Algodones Dunes that provided one of the clearest illustrations of the Center's use of aggressive negotiation to get more lasting protections than litigation alone. The Algodones Dunes in the southeastern corner of California is the largest area of active sand dunes in North America, covering 160,000 acres. Only about 25,000 acres were protected as wilder-

ness, and thus off limits to vehicles. The rest of the dunes had been proposed as wilderness, but that area was dropped as part of the political compromises leading to the California Desert Protection Act. In the meantime,
it became an increasingly popular spot for motorcycles, dune buggies, and
other off-road vehicles. By 1999, on Thanksgiving weekend alone, more
than a hundred thousand off-roaders converged on Algodones. All of this
vehicle activity damaged the fragile desert vegetation, including the Pierson's milk-vetch, a type of flower that had already been listed under the
ESA due to an earlier lawsuit by the Center.

The threat to the flower was one of the bases for the Center's litigation
against the BLM. Using a Section 7 claim, the Center could request that
all off-road vehicle use on the dunes be halted until the BLM underwent
consultation with the FWS over the effects on the flower, but it was also
quite possible that the judge might lift the injunction as soon as the BLM
began even a perfunctory consultation process. Therefore, the Center used
the threat of a total shutdown for the short term in order to negotiate for
more lasting protections. Since the edges of the dunes had received the
most off-road vehicle traffic, they were also the places where the desert
vegetation had already been largely eliminated. The Center was willing
to forgo a short-term injunction on these areas in order to get a much longer
injunction keeping vehicles out of about 50,000 acres in the heart of the
dunes, with the stipulation that this closure would stay in place until the
consultation process was competed and a management plan that addressed
the needs of the endangered species was finalized.

The Center's California desert case settlements were completed shortly
before President George W. Bush took office. The Bush administration
could not undo the desert case settlement, but it was strongly resistant to
the implementation, thus requiring that Cummings and an expanding roster of Center attorneys would have to return to court again and again in
order to get these settlements enforced sufficiently. Nonetheless, the heart
of the Algodones Dunes still remained closed to off-road vehicles at the
end of Bush's second term.

While the California desert case was the Center's biggest lawsuit in 2000,
it was only one example of the Center's activities that year. When the
Center prepared a month-by-month list of its achievements in 2000, the
various victories from the desert case represented only four of the eighty-
seven accomplishments on that list.[96] At the start of the new millennium,
the Center for Biological Diversity had moved well beyond its modest

beginnings in New Mexico and had an output in terms of ESA litigation, listings, and critical habitat designations that exceeded even the largest national organizations.

Continued Growth amid the George W. Bush Years

The arrival of the administration of George W. Bush in 2001 marked a significant shift in political landscape for endangered species protection. Bruce Babbitt was replaced by Gale Norton as secretary of the interior. Norton had previously been the top lawyer for James Watt, the notoriously antienvironmental secretary of the interior under Reagan, and then continued working with him at the Mountain States Legal Foundation, a "property-rights" advocacy group. While not romanticizing the Babbitt era, Center activists were keenly aware of the ways that the Bush administration presented new challenges for biodiversity protection. The Center would continue to win in court, but, as with the California desert case, the Bush administration's resistance hindered translating those court victories into on-the-ground protections. As Center attorney Brendan Cummings described,

> The Bush administration seems to have no problem repeatedly losing. They'd rather do something and be sued repeatedly and lose and then try it again, than lose once and then correct it so as to avoid more litigation. It's a funny thing; they say that we are using the court system for our policy purposes, but the Bush administration is very much using a court-based strategy as well. They consciously choose to implement decisions they know are illegal and figure that not all of them will be challenged. Those that are challenged, they'll likely lose on. But that still delays implementation of regulations they don't want to implement or keeps bad rules on the books for the couple of years it takes us to litigate them.

As a result, more of the Center's energy had to be put into litigating the same issues over and over again.

Another demand on the Center's resources was the proliferation of lawsuits filed by opponents over critical habitat designations. While economic considerations are supposed to play no role in listing decisions, once a species is listed, the Fish and Wildlife Service is required to take economic effects into account in determining its critical habitat. However, because the agency had contended that critical habitat did not provide significant additional protection over listing, it treated the economic impact of any critical habitat designation as negligible. As a result, these critical habitat designations were vulnerable to procedural lawsuits from ESA opponents who argued that the agency's economic assessments involved in those designations were inadequate. The Bush administration was quite willing to settle these suits by dramatically reducing the size of the critical habitat, so the Center was forced to intervene in these cases to prevent that outcome.

Species listings were another challenge. During George W. Bush's first term, the FWS only listed species when compelled to do so by the courts. Therefore, the Center continued to use litigation to keep the pressure on the administration. By the early 2000s, the Center was responsible for about half of all the wildlife-related cases handled by the government's attorneys.[97] In response, the Bush administration agreed to a "miniglobal settlement" with the Center, Biodiversity Legal Foundation, California Native Plant Society, and Southern Appalachian Biodiversity Project to have the FWS make listing decisions for twenty-nine species.

Beyond that settlement, two notable listings during Bush's first term—the Puget Sound population of the orca (killer whale) in Washington and the California tiger salamander—were indicative of the distinctive organizational dynamics of the Center. In each case, the listing petition was prepared by a Center attorney who had taken a personal interest in the fate of a species outside of the issues that she or he was initially funded to work on. Brent Plater, who was responsible for the orca listing, had been hired by the Center with funding from a Pew grant for trout protection. Kassie Siegel got the California tiger salamander protected while being funded to work on urban sprawl issues in Southern California. The California tiger salamander lives in habitat that is imperiled by urban sprawl, but not in the part of California targeted by Siegel's funder. Brendan Cummings contended that the salamander listing was "the most significant species listing of the first term of the Bush administration," and that its potential impact for challenging urban sprawl in central California could be comparable to the spotted owl's impact on logging. However, if the Center had required the activities of its staff to be entirely determined by its funders, these listings might not have happened. It was the Center's organizational flexibility that gave staff the latitude to pursue independent projects (along with the high productivity of the staff to accomplish these projects in addition to their funded work) that allowed these species to be protected.

During George W. Bush's first term, the Center also took on the listing cases of the Biodiversity Legal Foundation. BLF's director, Jasper Carlton, faced health problems, and a merger with the Center provided a way to ensure that work on protecting those species would continue. This merger brought the grassroots biodiversity litigation campaign full circle, with its founder's work being united with that of his most accomplished protégés.

The Center continued to grow in other directions as well. New offices opened in Alaska, Colorado, and Oregon, and the Center expanded into new areas of biodiversity protection and the use of new laws. For example, Brendan Cummings created the Center's oceans program and developed the first litigation under the High Seas Fishing Compliance Act of 1995. That suit, done with coplaintiff Sea Turtle Restoration Project, ultimately led to the shutdown of the longline fishery off California, a particularly destructive type of fishing that incidentally killed many endangered sea turtles, marine mammals, and seabirds. That case dramatically transformed the National Marine Fisheries Service's management of high sea fisheries beyond California by requiring environmental review and regulation of these previously unregulated fisheries. Cummings also began developing a litigation strategy for the issue of global warming, which would ultimately lead to the protection of the polar bear under the ESA (see chapter 6).

Center cofounder Peter Galvin developed an international program that explored innovative applications of U.S. laws to challenge environmentally damaging activities outside of the United States. One of the most notable international cases involved the Okinawa dugong in Japan. The dugong, a relative of the manatee (sea cow), is a marine mammal that grazes on sea grass along coastal waters. Okinawa is home to the most imperiled population of dugongs, with only about fifty individuals remaining. The island of Okinawa is also home to a high concentration of U.S. military bases, and the survival of the dugong was threatened by a proposed new Marine airbase to be built on top of a coral reef in prime dugong habitat where they feed and rear their young. Peter Galvin was invited by Japanese environmentalists to visit Okinawa to help protect the dugong from the military project. There he found that the dugong had such cultural significance to the Okinawan people that this species had been designated as the equivalent of a national landmark under Japan's Law for the Protection of Cultural Properties. Galvin then developed a legal strategy to challenge the proposed base under U.S. law, using the National Historic Preservation Act. This law included a provision that precluded the U.S. government from harming properties in another country that have been designated as the equivalent of national monuments or historic places by that country. In 2003, the Center, in conjunction with Japanese environmental groups, filed suit against Secretary of Defense Donald Rumsfeld to halt construction of the airbase under the requirements of the National Historic Preservation Act. It was the first litigation to ever use the international provisions of this law. The lawsuit ultimately prevailed in 2008.

In addition to its innovative legal strategy, another distinctive feature of the dugong case was its timing. The Center's lawsuit against Rumsfeld was filed at a time when American soldiers were engaged in combat in Afghanistan and were about to be sent to Iraq. Filed at the height of public support for the "War on Terror," a lawsuit that questioned actions by the U.S. military risked being labeled as unpatriotic. Galvin described being acutely aware of the potential for controversy around the airbase case but not letting it deter the Center from pursuing the protection of the dugong:

> Where other groups might have shied away from it, we didn't. And frankly I think the decision has served the issue well and it has served us well. Part of the Center's advancement over the years is that we are willing to take on issues that other people aren't willing to take on because there is an appearance that it is a politically difficult issue.

The Center was able to make progress on the dugong case despite the unfavorable political conditions of the early 2000s. At the same time, at an organizational level, the Center continued to grow despite unfavorable economic conditions. The stock market collapse in 2001 had a devastating effect on the resources available to grantmaking foundations. Around the same time, the Center lost some key allies in the foundation world when the Foundation for Deep Ecology phased out its U.S. grants program and Peter Bahouth departed from Turner Foundation. Yet, despite these developments, the Center did not experience the financial constriction suffered by many other grassroots groups at this time. It was able to avoid this problem by diversifying its funding sources.

For example, as the Center hired its own full-time staff attorneys, it was able to recover the attorneys' fees from the government in the cases won by its staff attorneys, which helped to make those positions less dependent on foundation funding. The fastest growing alternative to foundation funding came from the Center's members. By the early 2000s, the Center had grown large enough to overcome the barriers-to-entry that kept smaller grassroots groups from using direct mail fundraising efficiently (see chapter 2). The Center's membership program also benefited from the unusually high level of media coverage that it received, even though, unlike most environmental organizations, it did not have staff people who were specifically assigned to do publicity for the organization. Instead, its lawyers and campaigners were very effective at doing their own media outreach, and the Center generated more news stories each year than many of the largest national organizations.[98]

The Center was thus particularly well positioned to build membership as it hired new administrative staff to focus on this task. In 1999, when the Center began a concerted direct mail program, it had around 4,000 members. The direct mail response rate dropped following September 11, 2001, when the resulting political climate was less conducive to the Center's sharp criticism of the Bush administration, but the program grew rapidly in 2003 and 2004, and by the end of that year had about 13,000 members. By then, its annual budget exceeded $2 million.[99] So despite the change to a hostile Republican administration, the post-9/11 political climate, and the economic downturn, the first term of George W. Bush was marked by continued victories and growth for the Center.

Becoming a National:
Getting It Right by Doing It Wrong

The Center for Biological Diversity has become one of the most successful grassroots biodiversity groups, both in terms of environmental protection and organizational growth. By 2004, it had grown from very modest beginnings into an organization with a staff of more than thirty people, offices in six states, an impressive tally of successful environmental litigation, a sizable membership, and a multimillion dollar budget. It had a 90 percent success rate in its litigation and was responsible for the protection of 335 species and more than 43 million acres of critical habitat.[100] In terms of its relationship to the dominant national environmental organizations, it had gone from an early "never mind the nationals" approach to a "transform the nationals" approach with the Endangered Species Coalition to ultimately being on the cusp of becoming a national organization itself, in terms of its size and influence. In this concluding section, I examine how the Center achieved this level of success and then explore the implications of that success for its future.

Integral to the success of the Center has been its use of an outsider strategy. Attorney Jay Tutchton—who had worked for National Wildlife Federation prior to working with the Center—offered this comparison:

> One thing that the Center had going on for it early on is that they were not insiders. National Wildlife Federation in a lot of ways had a revolving door with the Fish and Wildlife Service. But the Center literally had a revolving door with no one. They had no inside access, no connections. They didn't care. So they were not capable of being pressured by insiders saying, "We'd like to do the good thing, but you don't understand our political problems." I think that was a strength.

The Center did not seek to appear reasonable in order to appeal to politicians or the media because it did not see them as the main source of its influence. Thus it was not readily deterred by controversy. It embodied cofounder Robin Silver's call for a "group that isn't afraid." As a result, the Center took the lead on pursuing contentious campaigns and litigation that more moderate groups would not initiate. And the biocentric perspective of the Center activists fostered particular attention to the protection of less charismatic, but ecologically important species that were often overlooked by the national environmental groups. The Center grew quickly within the spaces left vacant by the nationals. The external political conditions of the Clinton years were particularly conducive to this growth. The Center's outsider strategy allowed it to be a more outspoken critic of the Democratic administration and to push more aggressively for stronger environmental protections. As a result, the Center readily distinguished itself from national groups that were tied to an insider political strategy and did not want to lose access to the Clinton administration.

The Center's extensive use of litigation as a tactic is the group's defining characteristic. As Kieran Suckling noted, "Litigation is a great equalizer." Litigation allowed the Center to have a big influence despite its small size. Because of the citizen-suit provisions in the Endangered Species Act and other environmental laws, lawyers could take these cases pro bono and receive attorney's fees from the government upon victory. The early Center was very effective at finding pro bono attorneys for its cases.

At the same time, Center activists were willing to live on next to nothing. Therefore, they were able to have a significant impact with very little funding. These early experiences fostered an independent attitude within the Center. While the Center's founders became able fundraisers, they maintained an organizational resistance to letting funding become a moderating force.

Another key component of the Center's success was its unconventional staff operating within an unconventional organizational structure. The Center's founders and much of its early staff came from a background in radical environmental activism. They were considered too radical to be hired by moderate environmental groups and often lacked the professional credentials that the nationals expected. The Center provided a space for them to pursue their activism and there they proved to be highly effective. In particular, Earth First! supplied a pool of talented activists for the Center who were used to working with few resources and little formal structure. Earth First! also served as the movement culture for the developing Center. It had fostered the Center activists' confrontational outsider approach and their biocentric values. And the anarchic nature of Earth First! had nurtured a resistance to traditional forms of hierarchical organization among those activists, which in turn encouraged a decentralized organizational structure. As a result, Center activists and the lawyers who worked with them highlighted how "nimble" the Center was—its ability to act quickly in response to new opportunities without a long, bureaucratic decision-making process.

In the 2000s, the political climate became more unfavorable to the environmental movement with the election of George W. Bush and the dominance of antienvironmental Republicans in Congress. However, since the Center worked primarily through litigation rather than legislation or administrative action, it continued to make significant accomplishments in species protection. Economic conditions also became unfavorable during this time. Yet while foundation grants to environmental groups plummeted, the Center had reached a size at which donations from its individual members provided substantial alternate sources of funding, so it continued to grow dramatically during this period. In 2004, the Center for Biological Diversity was comparable to some national environmental organizations in terms of its size and scope. In terms of membership, its numbers that year were similar to those of Defenders of Wildlife around the time of the first Earth Day. And in terms of influence, the Center already had more of an impact on species protection through listings and critical habitat designations than even the largest nationals.

But what does it mean for the Center to become a national? After all, the national organizations are generally associated with qualities that Center activists have long criticized, such as a focus on political access and the resulting aversion to controversy and conflict. Does the organizational success of the Center mean that it, too, must take on these qualities? Other national organizations that were more aggressive in their early years ultimately developed a more moderate approach. Consider, for example, the experience of the Environmental Defense Fund. At its inception, EDF looked remarkably similar to the Center. It began in the late 1960s as a small group of aggressive lawyers and biologists, whose early mottos included "EDF: Sound Science, Racy Law" and founder Victor Yannacone's famous rallying cry, "Sue the bastards!"[101] Yet by the 1990s EDF had taken on a notably more conservative approach and was a frequent counterpoint to the Center on endangered species protection strategies.

Rather than assume that large environmental organizations inevitably become more moderate, it is useful to consider the specific factors that have led to these changes in other organizations in order to evaluate the extent to which they are applicable to the Center. I will focus on three factors: the role of the organization's founders in relation to the board of directors, the process of professionalization within the organization, and the extent to which the organization's activity focuses on lobbying in Washington, DC.

Founders and the Board of Directors

One factor identified by historian Stephen Fox in his account of the early environmental movement is the moderating effect that follows the loss of influence of an organization's founders, who are usually supplanted by the board of directors.[102] This pattern is also evident in more recent phases of the environmental movement as well. For example, journalist Mark Dowie described a key intervention by the Environmental Defense Fund's board of trustees against its founder in shaping its early development:

> It was litigation against these corporate violators that landed Victor Yannacone and his colleagues in deep trouble with their own trustees and funders, especially the Ford Foundation. They were simply too adversarial in their rhetoric, too confrontational in court, and too

successful against companies with close association to foundation trustees. The "sue the bastards" style had to go, and so did Yannacone. He was fired from EDF in 1970.[103]

Without this intervention by its board, EDF might have continued along a more confrontational path than the one it ultimately followed.

By comparison, the founders of the Center for Biological Diversity have remained firmly in control of their organization. Kieran Suckling, Peter Galvin, Robin Silver, and Todd Schulke have been the primary or sole board members throughout the group's history, as well as simultaneously being staff. Schulke declared, "Our board is probably singularly unique in the conservation world." The Center's board of founding staff is certainly unusual and runs contrary to conventional models of organizational management. Typically, board members are detached from the staff, have special connections to large funding opportunities, have professional credentials in management or law, and increase the overall diversity of the board. The founding board members of the Center did not have any of these characteristics. From the perspective of nonprofit management, the Center's board structure might seem like an example of what not to do. However, instead of seeing the Center's unconventional board as a problem, it may be an important asset.

To assess this point, it is helpful to distinguish between organizational maintenance (i.e., maintaining and expanding the organization's resources) and organizational mission (i.e., achieving the social change goals of the organization). A conventional board member is selected for characteristics that help to serve organizational maintenance, particularly by appealing to funders and providing new funding opportunities. Adding this type of board member may benefit the organization's growth, but not necessarily the organization's mission. Because ultimate authority over the organization lies in the hands of its board, the board of directors controls the strategic rudder of the organization. Expanding the board puts more hands on that rudder. Moreover, the type of board member who would contribute to fundraising will not necessarily share the same strategic vision as the founders and may be more averse to controversy. (Indeed, many national environmental organizations, including the Environmental Defense Fund, National Audubon Society, National Wildlife Federation, and World Wildlife Fund, have added board members from corporations that were directly implicated in environmental degradation.[104]) More hands on the strategic rudder brings a greater chance of going off course. Thus, a small, homogeneous board of founders can be an advantage in terms of preserving the organizational mission.[105] The Center's founders with their shared background in radical environmental activism have been unlikely to restrict other staff members from pursuing aggressive biodiversity protection actions that a more conventional organization's board might have deemed too controversial. They have done a particularly good job of maintaining the Center's contentious but effective outsider approach to litigation, even as the organization has grown from the margins to become an established presence in environmental policy making. The Center's unconventional board of directors has thus helped protect the organization from the constraining role that boards have played in other national organizations.

At the same time, the organizational maintenance of the Center has fared quite well even without the board focusing on that role. As Todd Schulke commented, "Our board does not have a fundraising function. Well, you know what? We're better off financially than almost any other group like us in the country. We haven't had the board play those traditional roles and it seems to have worked out just fine." Ironically, by prioritizing organizational mission, the board may have best served the Center's organizational maintenance as well.

Finally, having the board also be part of the staff has contributed to the nimbleness of the Center. Because those board members have been directly involved in the day-to-day campaigns, they could quickly respond to opportunities that a more detached board might miss. And the Center has retained this nimbleness despite extensive organizational growth. Longtime Center activist Shane Jimmerfield concluded that the founders' control of the board "has allowed the Center to maintain that small-group dynamic" amid this growth.

The Center's unconventional board structure has thus been an important factor in the success of the group while at the same time helping to protect it from some of the moderating pressures experienced by other national organizations.

Professionalization

In his chronicle of the American environmental movement, *Forcing the Spring*, environmental scholar Robert Gottlieb highlighted the professionalization of the national environmental organizations from the 1970s into the 1990s and the resultant moderating effect on those organizations. By the early 2000s, the Center was likewise experiencing a growing professionalization of its staff. As its budget grew, it could offer better pay to staff. In the early 2000s, the salaries were still comparatively low—about $30,000 for most staff—but they were nonetheless a marked shift from the poverty-level wages that had characterized the Center through much of the 1990s.[106] This shift also allowed the Center to attract new staff members with more specialized skills, such as in law, science, and fundraising.[107] However, this expansion brought with it a marketed shift in the background of its staff. While much of the staff hired in the 1990s had previously been involved Earth First!, few of the newer staff members with professional credentials shared this experience.[108]

While the Center was shifting to a more professionalized staff in the 2000s, that professionalization did not necessarily lead to moderating effects that would constrain the organization's effectiveness. One example of the possibility of professionalization without moderation can be found among some of the Center's staff attorneys. Some of these attorneys had roots in the Earth First! community. Some of them also sidestepped the increasing specialization within the Center. Rather than simply serving as the litigators for a campaign, they functioned as campaigners in their own right, initiating campaigns, teaching themselves the science around the issues, and doing their own media outreach. This arrangement might seem a bit counterintuitive; one might assume that it would be more efficient to have the lawyers focus specifically on the legal issues in which they were trained. Yet these lawyers' unconventional approach led to some of the Center's most notable accomplishments of the early 2000s, such as the California desert protection case, the tiger salamander listing, and the shutdown of longline fishing off the Pacific Coast. In these campaigns, the attorneys combined the Center's decentralized approach with professional skills in a manner that was unconventional but proved to be quite effective.[109]

Working in Washington, DC

Amid its organizational growth, the Center retained its underlying commitment to an outsider strategy for social change through direct pressure. This remains one of the central characteristics distinguishing the Center from the moderate national environmental organizations that rely on an insider strategy of access and compromise. In journalist Mark Dowie's analysis of the modern environmental movement, this insider approach became deeply entrenched in the national organizations as they shifted more of their focus to Washington, DC and access to politicians.[110] An organization that relied on this strategy was likely to moderate its own advocacy so as not to jeopardize its political access. It is not surprising, then, that many grassroots activists came to see a DC-focus as having a detrimental impact on environmental advocacy.

Yet by the early 2000s, the Center had staff member Brian Nowicki traveling frequently to Washington to work on federal policy issues, and after 2004 they added full-time staff based in DC. At first glance, this development might seem like evidence of the Center becoming more like the national environmental organizations. However, closer examination suggests that this may instead be yet another example of the Center reenvisioning characteristics of the nationals in an unconventional manner that avoided some of the associated pitfalls. Nowicki stressed that, despite the Center's greater involvement in DC policy work, "We don't want to become a DC-based group where our thoughts on policy and politics are based on what is happening inside the Beltway." Instead of focusing primarily on access to politicians, the Center's policy staff emphasized their work on influencing on the biggest environmental organizations there. As Kieran Suckling noted, "The most important people I lobby are the Sierra Club, The Wilderness Society, and those other groups in DC, not Congress.".

Sometimes this approach involved trying to get the nationals to not endorse compromise legislation that would substantially weaken endangered species protection, akin to the Center's intervention in the Endangered Species Coalition in the 1990s around the Environmental Defense Fund's legislation. In addition to preventing detrimental compromises, the Center's policy staff worked on encouraging the nationals to take stronger positions on biodiversity protection than they might have otherwise adopted. In this way, the Center's work in DC helped shift the terms of the ESA policy debate and infused more of a grassroots perspective in the efforts of the DC-focused environmental organizations.

At the same time, the Center's influence with those organizations did not depend on it appearing moderate and uncontroversial. Instead, it became an influential voice in that community, despite its smaller size, specifically because it had prevailed in contentious endangered species protection campaigns that most national organizations were reluctant to touch. As Suckling noted, "Once you do that enough and you're effective, you get a seat at the table, and it's not because people want to give you a seat at the table—it's because they have no choice." In this regard, the basis of the Center's influence in these circles stems from their continued use of hard-hitting litigation. Thus the Center's unconventional emphasis on lobbying the nationals rather than politicians may offer a way for the Center to participate in policy making in Washington, DC that sustains the group's bold approach to environmental advocacy rather than moderating it.

Navigating an Alternate Course

As the Center for Biological Diversity has grown, it has taken on some of the characteristics of the national environmental organizations that Center activists long criticized. However, in some key aspects—such as its unconventional approaches to its board structure, the role of professional staff, and its activity in Washington, DC—the Center has reenvisioned those characteristics in a manner that may help to shield it from some of the moderating influences that have constrained other national organizations. It remains to be seen whether the Center can successfully navigate an alternate course to becoming a national without those constraints, but so far its path looks promising.

Throughout George W. Bush's second term, the Center continued to experience rapid growth in the size of its staff, membership, and budget. By the end of 2008, it had reached a staff size of sixty-four people, with 60,000 members and an annual budget of more than $6 million. At the same time, it continued to undertake bold and influential biodiversity protection campaigns, including a new program on global warming (see chapter 6). In this regard, the success of the Center offers valuable lessons for the rest of the environmental movement. Its central lesson was best summarized by Kieran Suckling when he said, "You don't have to be moderate to be effective."

6

Boldness Has Genius: The Lessons of Grassroots Biodiversity Activism for the Campaign against Global Warming

In this concluding chapter, I return to the question of how small grassroots biodiversity groups with few resources were able to bring about an unprecedented increase in forest and endangered species protection in the United States during the last two decades. I contend that their success depended not only a powerful tactic—litigation—but also on a context wherein that tactic could be applied without significant constraint. Understanding how the grassroots biodiversity groups avoided constraints that snared the national organizations can provide valuable lessons for the environmental movement. It also raises questions about the centrality of the dominant national organizations for achieving environmental protection. I therefore reexamine the relationship between the grassroots and the nationals and consider its implications for funding distribution within the environmental movement. Finally, I conclude by exploring the lessons that the experiences of the grassroots biodiversity groups offer for the emerging campaign to address global warming.

Achieving Success by Avoiding Constraints

To understand the success of the grassroots biodiversity groups, I return to the six dimensions of social movement analysis that I introduced at the beginning of this book—tactics, strategy, movement culture, organization, funding, and political conditions.

Tactics, Strategy, and Movement Culture

Litigation has been the defining tactic of the grassroots biodiversity groups. That said, it was not the only tactic used by those groups. For example,

215

the Headwaters campaign involved mass civil disobedience, consumer boycotts, shareholder resolutions, and a variety of other tactics. Likewise, legislation was a central component of the zero-cut campaign. However, litigation was consistently the most influential tactic in all three case studies. It allowed the grassroots groups to enforce key environmental laws even when the government agencies charged with implementing them were resistant to doing so.

As we have seen, it is not sufficient simply to have strong environmental laws on the books. Laws are only effective when someone is willing to enforce them. While the national environmental organizations made some attempts to use litigation to get environmental laws enforced, their use of this tactic was constrained by their insider political strategy and the resultant aversion to controversies that might hurt their political access. In contrast, many of the grassroots groups were founded by activists who participated in or were influenced by Earth First! and had embraced its outsider strategy for social change. Because of their outsider strategy, these groups did not have to worry that controversial litigation would hurt their access to politicians. As Kieran Suckling of the Center for Biological Diversity noted, "Many groups are hampered by fear of upsetting their congressional connections, their funders, the media, et cetera. While we feel the pull of such things, we daily remind ourselves that social change comes with social tension and our job is to create that dynamic tension, regardless of the pressure to back down or compromise."[1] The grassroots biodiversity groups were therefore able to pursue forest and wildlife protection lawsuits that the national organizations were unable or unwilling to initiate.

Litigation was particularly important to the development of grassroots groups because court injunctions could have powerful impacts on environmental policies—for example, halting all logging on a national forest—without requiring the litigants to have the extensive resources or special political access of the large national environmental organizations.[2] In this context, the relative influence of the grassroots groups and the nationals shifted dramatically. As Kieran Suckling of the Center for Biological Diversity explained,

> Litigation is a great equalizer. The common view is of radical groups being these small voices in the wilderness and they're crying from the margins; maybe they help to keep people a little bit honest, but ultimately, they're not at the center of policy change. There are plenty

of examples of that. If you rely on being "a voice," then that's probably going to happen. It's extraordinarily difficult to force a voice into the mainstream, because how do you compete with your little voice against the larger voices of the opposition and the big environmental groups? However, once you start to litigate, it's really the great equalizer because at that point you're not simply saying, "Listen to my voice, please"; you're saying, "I'm going to harness the voice of a federal judge, and that does absolutely have to be listened to." And so, if you're smart, with relatively little resources you can bring litigation that actually changes the world.

An outsider strategy allowed the grassroots groups to pursue litigation more fully than the nationals, and the success of that tactic enabled the grassroots groups to maintain their outsider strategy rather than seeking more conventional forms of influence through an insider strategy. In this way, grassroots groups were able to circumvent the constraints that the insider strategy had imposed on the national organizations.

While the benefits of the outsider approach are evident with hindsight, the use of an outsider strategy is a fragile accomplishment. It goes against powerful cultural codes that channel people into trying to make social change only through conventional forms of political participation. Groups using an outsider strategy are regularly involved in conflicts that lead to controversy and condemnation from prominent figures in the government and media. As many of the grassroots activists noted, people want to be liked, and as a result, most people tend to go along with detrimental compromises rather be seen as difficult or disruptive. It is not easy to adopt or sustain an outsider strategy. Activists who choose to pursue this strategy often need the support of a radical movement culture to counterbalance these societal pressures.

Earth First!'s radical movement culture was particularly important to the development of the grassroots biodiversity groups.[3] It nurtured a generation of activists who were skeptical of compromise, and who did not see conventional forms of political participation as the sole means of protecting the environment. Some of these activists started new grassroots biodiversity groups such as the Center for Biological Diversity. Earth First! also provided a pool of radicalized activists who went on to work for groups such as the Environmental Protection Information Center (EPIC). And other groups such as the John Muir Sierrans were able to reach a national audience and find new allies through the open forums provided by the

Earth First! Journal and *Wild Earth.* The radicalizing effect of Earth First!'s movement culture was a key reason why there was a proliferation of bold new biodiversity groups in the late 1980s and 1990s.[4]

In short, litigation provided a powerful tactic and an outsider strategy— nurtured by a radical movement culture—provided the willingness to use that tactic more fully, which ultimately enabled the grassroots groups to have an unprecedented level of influence on federal biodiversity protection policies.

Organization and Funding

While litigation tactics, outsider strategy, and radical movement culture played mutually reinforcing roles in the success of the grassroots groups, organization and funding played more problematic roles. They could both enable and constrain the work of biodiversity groups. Consider, for example, the role of organization. Some level of formal organization beyond the informal structure of Earth First! was advantageous to biodiversity activists. The new grassroots groups were able to receive funding and provide activists with modest incomes. The formation of these groups allowed radical activists who otherwise would have been unable or unwilling to be employed by the more moderate national organizations to work on biodiversity protection full time. In contrast, the John Muir Sierrans' reform efforts within the Sierra Club were hindered because they were not able to create a formal organization due to concerns about being sued by the Club and thus could not receive grants to support their work. (That said, some John Muir Sierrans were able to sustain their activism by working for grassroots organizations such as the John Muir Project, Heartwood, and Native Forest Council.)

While formal organization offered some benefits over informal organization, bigger was not necessarily better. Organizational growth could be a moderating force, causing groups to become more risk averse and less willing to pursue contentious actions.[5] This tendency was particularly evident in the behavior of the largest national organizations. The nationals were already constrained by their use of an insider strategy for social change, but this cautiousness was further reinforced by the bureaucratic conservatism that had accompanied their organizational expansion. By comparison, smaller groups could be more nimble and bold in their actions. Ironically, the success and subsequent growth of these groups could put a damper on their effectiveness—although in the case of the Center for Biological Diversity, unconventional organizational arrangements, including a board of directors made up of founding staff, helped to maintain a small-group dynamic amid rapid expansion. Nonetheless, the constraints associated with organizational growth represented an ongoing challenge for grassroots biodiversity groups.

External funding, particularly from grantmaking foundations, also played a complicated role for the grassroots biodiversity groups. On the one hand, foundation funding allowed these groups to do their work. On the other hand, there was also concern among these groups about foundations being a moderating force. As Tim Hermach of the Native Forest Council stated, "Foundation money behind a compromise position tempts nonprofits to moderate their hard-line stance or risk being left out in the cold."[6] Examples of foundations acting in this way were evident in the case studies in the previous chapters. Pew Charitable Trusts was prominently associated with this concern, most overtly in its intervention in the Endangered Species Coalition. Likewise, grassroots forest groups worried that Pew and similar foundations would not support them if they endorsed a zero-cut position. In other cases, well-meaning interventions by foundations that were intended to aid the grassroots biodiversity groups through professionalization (such as the Training Resources for the Environmental Community process, noted in chapter 5) and consolidation in formal coalitions (such as the creation of the Headwaters Forest Coordinating Committee) inadvertently risked undermining the distinctive qualities that had made these groups so effective.

Nonetheless, while foundations sometimes played a problematic role, foundation funding was indispensable for the development of the grassroots biodiversity groups. These groups faced a situation in which forest and wildlife protection advocacy was dominated by the national environmental organizations. Those organizations had grown rapidly in the 1970s and 1980s through direct mail fundraising. However, the organizational success of the nationals created barriers to entry for new groups seeking to follow this fundraising path. Small new biodiversity protection groups had little hope of sustaining themselves through membership donations. They did not have the name recognition of the nationals and thus were unlikely to be selected by members of the public who also received many direct mail appeals from the better-known organizations. Moreover, the economies of scale for direct mail fund raising made it prohibitively expensive for most grassroots groups.[7] Foundations provided a crucial alternative source of support for the new grassroots groups.

However, it should be noted that the start-up funding for these groups came primarily from a small subset of maverick foundations, rather than the larger and more moderate environmental grantmakers. What distinguished many of the maverick foundations was that their staff had previously been activists in environmental groups that used an outsider strategy, such as Earth First! (Foundation for Deep Ecology, Levinson Foundation), Greenpeace (Turner Foundation), and Earth Island Institute (Patagonia).[8] These funders were also less likely to impose constraints on the activity of the grassroots groups. Thus the new groups were able to avoid the fate of earlier litigation organizations such as the Environmental Defense Fund, Natural Resources Defense Council, and Sierra Club Legal Defense Fund, which the Ford Foundation's oversight had steered away from more radical approaches early on.

Likewise, the backgrounds of the grassroots biodiversity activists helped to limit funder influence over them. Many had begun as unpaid volunteers in groups such as Earth First!. They did not expect to receive a salary in order to engage in activism and they were used to accomplishing a lot with minimal resources. Furthermore, Earth First!'s movement culture had nurtured an antiauthoritarian attitude among them. As a result, they were less likely to be tempted to significantly alter their advocacy work to appease a funder. As attorney Jay Tutchton recounted, "Whereas some groups had a huge budget and a lot of people dependent on that, because the Center for Biological Diversity came up from no budget, they didn't really care about funder influence. If they pissed off a funder, 'So what?' was the view of the Center." Thus the smaller size and the volunteer background of the grassroots groups gave them greater autonomy in relation to foundations. It allowed them to subsist on the support of a handful of maverick funders without significantly moderating their approach.

At the same time, dependence on a small subset of foundations left the grassroots biodiversity groups in a vulnerable position. Foundation support could decline rapidly in concert with the stock market, with the grassroots groups often hardest hit by the funding contraction. Foundations could also replace key staff, as happened with Peter Bahouth of Turner Foundation. And foundations could shift their funding to other issues, as happened with many of the Headwaters grantmakers following the Headwaters deal, as well as the Foundation for Deep Ecology and Levinson Foundation in relation to national forest protection. All three of these changes happened simultaneously in the early 2000s, leading to economic hardship for most grassroots biodiversity groups. The most successful groups during this time were able to develop alternate sources of support, such as the Center for Biological Diversity's membership drive, but on the whole, the vagaries of foundation funding hindered grassroots groups in the early 2000s.

In total, grassroots biodiversity groups had a multifaceted relationship with foundations. On the one hand, the new groups certainly benefited from receiving foundation funding in order to pursue their advocacy work full time, while overcoming some of the fund-raising barriers to entry that generally favored the existing national environmental organizations. On the other hand, foundation funding could impose new forms of control over groups and serve as a moderating force. The grassroots biodiversity groups were most successful when they limited the amount of control that foundations exercised over them. They did so by operating on modest budgets, having an antiauthoritarian attitude toward funder influence, relying primarily on support from maverick foundations with staff from similar backgrounds, and, where possible, finding alternate resources beyond foundation grants.

Political Conditions

Like organization and funding, political conditions played a multifaceted role in the success of the grassroots biodiversity groups. Conventionally, the Democratic Party has been seen as an ally of environmentalists, while the Republican Party has increasingly become dominated by opponents of environmental protections. One might assume therefore that election of a Democratic president would lead to more biodiversity protection, while a Republican White House would result in less. However, the experiences of the grassroots biodiversity campaigns tell a more complicated story.

For example, contrary to those assumptions, grassroots biodiversity groups won important victories under George H. W. Bush. For forest activists, the Bush administration's "train wreck" approach to the northern spotted owl inadvertently resulted in an unprecedented injunction that shut down national forest logging throughout the Pacific Northwest. Furthermore, the Bush administration settled Jasper Carlton's massive endangered species listing lawsuit, which created the conditions for the subsequent surge in listings during the 1990s. Even during the much more hostile George W. Bush administration, grassroots groups still had notable victories. For example, grassroots forest activists continued to push national forest logging levels down even lower than they had been during the

Clinton administration. Overall, biodiversity activists were able to make some important advances during Republican administrations.

Likewise, the arrival of a Democratic president did not automatically offer salvation for biodiversity protection efforts. Indeed, the Clinton administration proved to be a disappointment on this front. With regard to national forests, Clinton ended the logging injunction in the Pacific Northwest by enacting a forest plan that allowed the continued decline of the spotted owl and that was opposed by many scientists. Moreover, Clinton forced forest protection groups to accept this plan and to release old-growth areas for logging by threatening to exempt the new plan from legal challenges. Clinton also signed off on the Salvage Rider, which led to a period of "logging without laws" on national forests.

Regarding endangered species protection, Clinton's secretary of the interior Bruce Babbitt was a strong proponent of habitat conservation plans, which often functioned as de facto exemptions from Endangered Species Act enforcement. Under Babbitt, habitat conservation plans were issued in unprecedented numbers and were combined with a No Surprises policy that locked in these exemptions for decades. Furthermore, Babbitt instituted policy changes in 1995 that made it more difficult for the public to get imperiled species protected. As a result, listing rates declined steadily after 1996. Then, at the end of Clinton's second term, U.S. Fish and Wildlife Service Director Jamie Clark instituted a moratorium on new listings of endangered species, even though Congress had not mandated doing so.

In the case of Headwaters Forest, although the Clinton administration negotiated purchase of the main grove, it did so in the context of a deal that did not cover enough of land to protect the forest ecosystem, that paid Maxxam an excessive price for that land, and that was tied to a habitat conservation plan that increased Maxxam's logging in the rest of the forest. Moreover, no deal would have been needed if the Clinton administration had been willing to enforce the Endangered Species Act with regard to Headwaters. Nonetheless, the Headwaters deal was touted as an environmental accomplishment for Clinton.

Other Clinton administration actions that were celebrated as victories for environmental protection similarly proved to be of questionable benefit. While the Clinton Roadless Rule offered incomplete but nonetheless welcome protection for parts of the national forests, the overall effect of that policy on logging levels was negligible. Likewise, natural areas designated as national monuments by Clinton were established in a manner that provided less protection than previous monuments. For example, when Clinton created the Giant Sequoia National Monument to protect some of the largest trees on earth, he did not give management authority to the National Park Service, as was the case with most other national monuments. Instead it would be controlled by the U.S. Forest Service, which continued to promote logging within the monument. In such cases, Clinton's environmental initiatives seemed more symbolic than substantive.

Nonetheless, the national environmental organizations often trumpeted Clinton's actions as great victories. This response was indicative of the problematic role of the nationals during the Clinton presidency. As Earth First! activist Judi Bari observed, "The problem I see with the Clinton administration is a relaxing of vigilance of some of the environmental groups and a belief that now the government is on their side. I don't think that the government is on our side."[9] At the same time, the national organizations that depended on an insider strategy were more reluctant to criticize a Democratic administration for fear of hurting their access to Democratic politicians. Overall, the environmental advocacy of the nationals became more muffled during Clinton's presidency.

The grassroots biodiversity groups did not have this problem. They did not use an insider strategy and did not have to maintain their access to Democratic politicians. They were thus freer to publicly criticize the actions (and inactions) of the Clinton administration that were detrimental to the environment. They were also more willing to pursue litigation against the Clinton administration for not enforcing environmental laws. As such, the grassroots groups came across as bolder and more consistent advocates for the environment than the nationals. Therefore, the Clinton administration was inadvertently beneficial to the grassroots biodiversity groups because those groups could better distinguish themselves from the nationals during this time. In contrast, during George W. Bush's administration, when the nationals were more willing to criticize the administration and both the nationals and grassroots spent much of their time on defense against attacks on environmental laws, the distinction between them was blurred.

For all of its shortcomings, the Clinton years were a period when more progress was made on forest and endangered species protection compared with the George W. Bush administration that followed. For example, during the Clinton presidency, grassroots biodiversity groups would often have to sue the government to get a species listed, but once the groups prevailed in the lawsuit, the agencies would generally implement the necessary protections. By comparison, under George W. Bush, ongoing resistance by the administration and agencies to implementing environmental protections, even after the grassroots groups had won in court, made progress much slower. In this regard, having a Democrat in the White House ostensibly did present a more favorable climate for environmental protection, but only because there were grassroots biodiversity groups that were willing to file the listing petitions, timber sale appeals, and litigation to force the Clinton administration to enforce existing environmental laws.[10]

The experiences of the grassroots biodiversity campaigns suggest that the environmental movement should take a more skeptical approach to its relationship with the Democratic Party. Chad Hanson of the John Muir Project warned that environmentalists should "not put everything behind one political party and then capitulate [on environmental advocacy] because we think that we have to in order to make sure that the Democrats stay in power." Having a Democratic president in and of itself does not ensure significant biodiversity protection. Moreover, during a Democratic administration, the dominant insider organizations are likely to take a less confrontational approach, which can constrain advocacy and result in more limited protections than otherwise might have been achieved. It is only when there are outsider groups willing to vigorously push a Democratic administration toward stronger biodiversity protection that the relative advantage of having a Democrat in the White House is realized.

Rethinking the Relationship between the Grassroots and the Nationals

The combination of effective tactics, enabled by an outsider strategy and a radical movement culture, and their relative success in avoiding the constraints connected to foundation funding, organizational growth, and electoral politics allowed the grassroots biodiversity groups to have a significant direct influence on federal environmental policies, despite their small size and limited resources. The successes of the grassroots biodiversity groups challenge conventional notions of the relationship between radical and moderate groups within a social movement. Prior to the proliferation of grassroots biodiversity groups, analyses of the radical and moderate wings of the environmental movement highlighted Earth First! as the primary example of a radical group. However, because Earth First!'s direct action tactics generally did not have much direct influence on environmental policy implementation, analysts focused on the question of whether Earth First! helped or hurt the influence of the moderate national environmental organizations. Critics of Earth First! feared that its controversial actions would harm the credibility of moderate environmentalists and thus harm environmental protection efforts. Others saw Earth First! as indirectly helping the moderate national organizations. For example, one group of policy scholars concluded, "The extreme positions of radical environmentalists . . . may make government and business leaders more likely to negotiate with 'moderate,' mainstream groups."[11] In either case, such analyses were predicated on the assumption that the moderate insider groups were the primary source of policy influence by the environmental movement.

However, the rise of the grassroots biodiversity groups expanded the potential types of relationships between moderate and radical groups. The new groups used an outsider strategy but still had a substantial direct influence on environmental policies. Therefore, it was no longer sufficient to simply ask whether radicals helped or hurt the effectiveness of moderate groups. Instead, there were three possible relationships between the moderate national organizations and the more radical grassroots biodiversity groups: the grassroots groups could hinder the effectiveness of the nationals; the grassroots and nationals could have complementary roles that boosted their overall effectiveness; or the nationals could hinder the effectiveness of the grassroots groups. I will now examine how well each of these relationship models fit the experience of the biodiversity protection campaigns discussed in the previous chapters.

The "counterproductive grassroots" perspective contends that grassroots activists who make strong demands for biodiversity protection can hurt the credibility of moderate environmental organizations, reducing the influence of the moderate groups and resulting in less environmental protection. An example of this perspective was the concern of former Sierra Club Executive Director Michael Fischer that if the Club endorsed a zero-cut position, "the club lobbyists would lose the access [the Sierra Club] has in Washington. Committee staffers would simply refuse to meet with them because their own credibility would be affected if they did."[12] However, after the Club's membership voted to endorse zero cut, such predictions did not prove correct. Indeed former Sierra Club board member Chad Hanson noted, "Everyone has seen and acknowledged since then that this criticism was dead wrong. In fact, not only was it wrong, but exactly the opposite was true. Everything shifted positively in our direction. More was possible. Bigger things were possible. More things could be talked about." By the early 2000s, zero-cut activists had played a central role in bringing logging on national forests down to its lowest level since the 1930s.

A review of the logging levels on national forests over the past three decades shows that this unprecedented drop in logging levels accompanied the arrival of the more radical grassroots forest protection groups in the late 1980s and 1990s (see figure 4.1). The laws that these groups used to stop the logging had been on the books since the 1970s, yet it was not until these new groups formed that the laws were more fully enforced. A similar pattern is evident with regard to the protection of wildlife under the Endangered Species Act. Following the formation of new grassroots biodiversity groups in the late 1980s and 1990s, an unprecedented number of imperiled species were protected under the ESA in the 1990s (see figure 5.1), followed by an unprecedented number of critical habitat designations in the 2000s (see figure 5.2). Both the listings and the critical habitat designations were the direct result of litigation coming primarily from the new grassroots groups. In short, the proliferation of more radical grassroots biodiversity groups in the late 1980s and 1990s corresponded to a marked increase in forest and endangered species protection. Such an outcome was the opposite of what a "counterproductive grassroots" relationship should produce.

The increase in biodiversity protection that accompanied the arrival of the new grassroots groups was instead indicative of either a "complementary" relationship or a "counterproductive nationals" relationship. In the former, the nationals and grassroots groups would both enhance each other's effectiveness, resulting in greater levels of protection than would have been achieved without the involvement of both types of groups. However, the greater levels of biodiversity protection achieved following the proliferation of grassroots groups could also occur in a "counterproductive nationals" relationship, with the actual increase masking an even higher level of protection that potentially could have been achieved by the grassroots in the absence of the nationals. It is, of course, more difficult to demonstrate this latter relationship because there has never been a time when there were grassroots biodiversity groups but no nationals. To assess which of these two relationship models fits best, it is necessary then to look at specific experiences of the biodiversity protection campaigns.

In the preceding three case studies, there was surprisingly little evidence of a complementary relationship. In particular, it was difficult to see the beneficial contributions of the nationals in terms of either litigation or legislation. The work of bringing down logging levels and increasing endangered species protection was done primarily through litigation. Much of this litigation was undertaken largely at the initiative of grassroots biodiversity groups, with the more moderate national organizations playing little or no role. At times, the nationals directly tried to prevent grassroots litigation. And even on those occasions when a national organization would sign on as a coplaintiff to a lawsuit, the impetus for the litigation generally came from the grassroots. Grassroots groups invited national organizations to join as coplaintiffs because the name recognition of the nationals was thought to contribute to gaining greater media attention for the lawsuit and greater stature for the litigation before the court. Yet grassroots activists complained that they frequently got little or no assistance from the well-funded national organizations in preparing these lawsuits. Moreover, when the litigation prevailed, the nationals were in a stronger position to claim credit in the media and to fundraise off the victory at the expense of the grassroots groups that had actually initiated the lawsuit.

That said, a complementary relationship does not require that the moderate national organizations participate in litigation. There could instead be a complementary division of labor in which the political efforts of the nationals resulted in the legislation that provided the basis for grassroots litigation. However, in actuality, this litigation was usually based not on recent legislation enacted at the behest of the nationals, but rather on laws that had been passed in the early 1970s, in which the dominant national organizations had played little role or even a counterproductive role (see appendix).[13] Instead, the main benefit from the legislative work of the nationals came when those organizations lobbied to defeat new antienvironmental legislation that would undermine the statutory basis for grassroots litigation. However, frequently the nationals' response to these legislative threats was more problematic. As the nationals' insider influence faded during the 1990s in the face of an increasingly hostile Congress, these organizations became more willing to engage in political dealmaking, which resulted in legislation that sacrificed some existing biodiversity protection provisions. Grassroots activists frequently expressed frustration with the shortcomings of the nationals' political efforts. Kieran Suckling of the Center for Biological Diversity said, "They weren't holding up what I believe was an implicit contract that the local and the regional groups do the on-the-ground stuff and the big groups do the political lifting and represent us in some manner [in Washington, DC]."

Furthermore, grassroots groups learned that they did not need to depend on the national organizations to introduce and advance environmental legislation, such as with the Forests Forever ballot initiative and Rep. Hamburg's bill for the Headwaters campaign; the National Forest Protection and Restoration Act of the zero-cut campaign; and the Center for Biological Diversity's intervention in the Endangered Species Coalition, which ultimately lead to a stronger Endangered Species Act reauthorization bill. When grassroots groups were able to get their own bills introduced (often over the objections of the nationals), while they did not have notably more (or less) success than the nationals in getting major legislation enacted, their stronger forest and endangered species protection bills helped to reframe those issues in a way that advanced their campaigns. Therefore, the legislative role of the nationals not only seemed counterproductive at times, but also unnecessary.

To be clear, the role of the nationals was not counterproductive simply because those organizations took an incremental approach. Indeed, the campaigns profiled in this book refute the conventional image of the radical grassroots groups as uncompromising idealists who are unable to accept incremental steps and thus must depend on the moderate national organizations to broker the deals that will advance biodiversity protection step by step. Grassroots groups did engage in negotiation, especially as part of the litigation process, and they accepted incremental steps that were not total victory. However, there were fundamental differences between the nationals and grassroots groups in their approach to negotiation and incremental steps that reflected their respective underlying strategies for social change. Grassroots groups would enter negotiations from a position of strength with the prospect of a court injunction against the environmentally destructive activity, whereas insider organizations that were less likely to litigate had comparatively little leverage in negotiations. As a result, the nationals were willing to endorse compromises that not only accomplished less than might otherwise have been possible, but also made it more difficult to achieve additional incremental progress afterward. By comparison, grassroots groups were in a stronger position to forgo compromises in which short-term benefits came at the expense of long-term progress.

The different approaches of the grassroots and nationals to incrementalism were evident in their respective responses to President Clinton's roadless areas proposal. Many grassroots zero-cut activists were willing to participate in securing roadless areas protections, but they did so in a manner that could also advance the broader goal of ending commercial logging on the rest of the national forest system. In contrast, the nationals limited their advocacy largely to what the Clinton administration had proposed, and they approached the roadless campaign in a narrow manner that ultimately made it more difficult for forests activists to pursue further protections beyond the roadless areas (see chapter 4).

This is not to suggest that the national environmental organizations deliberately sought to play a counterproductive role in biodiversity protection campaigns. Instead, that outcome was an inadvertent result of their commitment to an insider strategy for social change. Reliance on an insider strategy limited these activists' vision of what could be achieved to a more narrow interpretation of political realities, making them more likely to settle for less. For example, when reflecting on her support for the Headwaters deal, Jill Ratner of Rose Foundation said, "there is a tendency for insiders to focus on what can we get done here and outsiders to be more focused on what are the needs of the community or what are the ultimate stakes. . . . To the extent that I screwed up in the way I did my work on Headwaters, it was because I was too limited by what I thought was attainable in thinking about what was necessary." The national organizations that worked primarily through an insider strategy were more likely to accept flawed legislation that provided only partial protections or even reduced some protections because they saw the alternative as being no legislation and no environmental protection. In contrast, outsider groups were not dependent on new legislation as the only way to expand biodiversity protection; they had found that they could more fully implement existing laws through enforcement litigation to achieve the same result.

Furthermore, the insider strategy of the nationals left them especially willing to accept less when Democratic politicians were promoting the compromise because they were reluctant to put too much pressure on their relationship with those politicians. For example, the nationals were unwilling to challenge the Clinton administration's Option 9 plan for the national forests of the Pacific Northwest, despite the objections to the plan from the scientific community. In addition, when limited environmental initiatives were enacted by Democratic politicians—such as the

Roadless Rule or the Headwaters deal—insider organizations such as The Wilderness Society were likely to laud those measures in the media. Their public celebration of the politicians who ratified these measures was an essential part of the quid pro quo of the insider strategy. However, their uncritical praise for these measures had a demobilizing effect that made it harder for environmental activists to achieve further progress. Grassroots forest activists subsequently faced the hurdle of having the media, public, and funders believe that the national forests and Headwaters Forest were saved. As a result, funding dried up, public participation declined, and it became difficult to get media coverage of these issues.

At other times, the insider strategy of the nationals caused them to hinder environmental protection even more directly. In legislative negotiations, environmental organizations relying on an insider strategy had to make concessions as part of their deal-making. Because the nationals were less involved in using litigation and at times saw lawsuits as a source of unwelcome controversy, they were more readily willing to trade away the legal tools that grassroots activists were using effectively to stop logging and protect endangered species. Thus, grassroots groups felt compelled to intercede to prevent the passage of detrimental compromises, such as the Center for Biological Diversity's intervention against Environmental Defense Fund's legislation that would have weakened the Endangered Species Act. When the grassroots were not able to intervene successfully, the resulting legislative actions supported by the nationals could be counterproductive for their campaigns, such as the role of the Sierra Club and The Wilderness Society in the legislation that set the groundwork for the Healthy Forests Restoration Act, and the role of the Club in the state legislation that funded for the Headwaters deal. The core problem with this sort of compromise was summed up by grassroots activist Ken Miller, who described the Sierra Club's role in enacting the Headwaters deal as "taking away our weaponry."

Even when the nationals did not play an actively counterproductive role, could the resources used by the nationals have achieved more if those resources had instead been put in the hands of grassroots groups? The grassroots groups described in the case studies accomplished much more in terms of timber sales stopped and endangered species protected per dollar spent compared to the nationals, and yet the nationals received the vast majority of the funding. Journalist Mark Dowie reported that the twenty-five largest environmental organizations were the recipients of 70 percent of all environmental funding, with the remaining 30 percent spread thinly among about 10,000 smaller groups.[14] In response, he proposed that those funds be redistributed so that the smaller environmental groups and the twenty-five biggest national organization each receive 50 percent of the total funding.[15] Given the successes of the grassroots groups, an increase in resources available to the grassroots would likely enhance the overall effectiveness of the environmental movement. However, Dowie's proposal was penned in the early 1990s before the level of direct influence achieved by the grassroots biodiversity groups was fully apparent. Based on the subsequent success of the grassroots groups, it is not clear that the role of the nationals warrants 50 percent of the total funding. In light of the developments during the last two decades, the environmental movement would benefit from a collective reassessment of the role of the large insider organizations in achieving environmental protection. This reassessment would need to address two key questions.

First, how much of the environmental movement's overall budget should go into DC-focused political advocacy? While most environmentalists see benefits to having some level of presence in the Capitol, it is not clear that the current concentration of resources there is justified. As Bryan Bird—a grassroots forest activist who had recently worked for a national organization—summarized, "We need a voice in Washington, but I don't know that we need hundreds of staff operating in Washington."

Second, should the resources that are to be spent on a Washington presence go to the existing national organizations that have relied on an insider strategy, or should that political advocacy instead be undertaken by grassroots groups that engage with federal policy-making processes but retain an outsider strategy? Grassroots activists expressed strong frustration with the lack of accomplishments from the dominant insider organizations. Mike Roselle—a cofounder of Earth First! who later worked in Washington—offered this assessment of the DC-based national organizations: "Those people cannot justify the amount of money we spend keeping them there. Send them home and let's start over. The money has got to go somewhere else."

Large-scale redistribution of resources along the lines that Roselle suggested would require a shift in priorities by funders, including foundations and individual "checkbook" members. They would need to channel their resources toward the groups that are most unconstrained in their environmental advocacy. Often this would mean moving away from sponsoring the well-known national organizations. While the national organizations are more likely to be invited by politicians to environmental bill signings and ribbon cuttings for new protected areas, their privileged access does not necessarily mean that those organizations were the primary force in achieving that outcome. Indeed, those insider ties are more likely to constrain the organizations' advocacy. While the national organizations are more likely to have professionally produced grant proposals and glossy magazines, funders should not assume that a more professionalized group is more effective. While professionalization is not inherently incompatible with environmental advocacy, the more professionalized groups have often become more moderate. Likewise, while being small does not guarantee taking a bold approach, in general, funders should be wary of very large organizations (and formal campaign coalitions) and the moderating effect of their institutional bureaucracies. If recent history is a guide, foundations and individual donors may accomplish more environmental protection per dollar by focusing their resources on bold grassroots groups.[16] As Mike Roselle summarized, "Most all of the victories over the last twenty years have come from the scrappy little campaigns."

Grassroots Activism and Global Warming

The case studies in the previous chapters focused on the period from 1989 to 2004. In the years immediately following 2004, there has been a dramatic increase in public concern over the issue of global warming. The growing levels of carbon dioxide and other heat-trapping gases emitted by fossil fuel burning and other human activities are raising the global temperature, resulting in droughts, hurricanes, flooding, and other extreme weather. This climate crisis not only threatens human health and survival, but also the survival of global biodiversity. Scientists have determined that more than one-third of all species of life on Earth may be committed to extinction by the year 2050 due to the effects of global warming if current levels of greenhouse gas emissions continue.[17] Global warming is likely to be the defining environmental issue of the twenty-first century. The challenge of addressing this vast problem is certainly daunting. To face this challenge, we would benefit from looking at what the environmental movement has done well in regard to other environmental issues. In particular, the success of grassroots groups in protecting forests and endangered species has been one of the U.S. environmental movement's major accomplishments of the past two decades. I will therefore conclude by exploring what lessons these biodiversity campaigns can offer for the emerging campaign against global warming.

As was the case with the biodiversity campaigns, there are already serious concerns being raised about the role of the dominant national organizations in addressing the climate crisis. Mike Roselle offered this assessment: "There's no leadership coming from them. They're just following. The proposals they are putting forward are no longer visionary." The nationals once again appear to be constrained by their insider strategy. Journalist Ross Gelbspan wrote, "Some of the country's leading national environmental groups are promoting limits for future atmospheric levels that are the best they think they can negotiate. Although those carbon levels may be politically realistic, they would likely be climatically catastrophic."[18] The constraints on the nationals limit not only their goals, but also the mechanisms that they propose for addressing global warming. Gelbspan observed that "a majority of climate groups—shackled by the cautiousness of their conservative funders and afraid of appearing too radical to the business and political establishment—continue to promote market-based approaches as a way to propel a global transition to clean energy."[19] However, their focus on promoting market-based carbon-trading systems that allow businesses to buy and sell the right to emit greenhouse gases faces growing criticism from grassroots activists. For example, former John Muir Sierran Michael Dorsey wrote in a newspaper editorial that "early evidence suggests that such a scheme may be a Faustian bargain . . . carbon trading has not resulted in an overall decline of the [European Union's] carbon dioxide emissions. Worse, the early evidence suggested that the trading scheme financially rewarded companies—mainly petroleum, natural gas, and electricity generators—that disproportionately emit carbon dioxide."[20] Such concerns led a group of grassroots activists to temporarily take over the Environmental Defense Fund's office in Washington, DC in protest over EDF's leading role in the promotion of carbon trading.[21] The focus on carbon trading has also affected the legislative strategies of the national organizations. As Brendan Cummings of the Center for Biological Diversity observed, "The nationals are so obsessed with getting a carbon trading bill passed—so as to be able to claim success that they've passed climate legislation—that they will support any bill that is seen as politically feasible, even if it is worse than what we have now, even if it guts powerful provisions of the Clean Air Act that we can currently use to regulate greenhouse gas emissions, and even if it sets us no further down the path to actually dealing with the climate crisis."

Beyond such criticism of the role of the nationals, grassroots biodiversity activists have also played a proactive role in addressing global warming. For example, forests have an essential role in absorbing carbon out of the atmosphere and thereby lowering greenhouse gas levels. By the same token, logging and other forms of deforestation release carbon, making those activities the second-largest source of carbon emissions globally.[22] As Chad Hanson of the John Muir Project noted, "You cannot meaningfully address climate change by just addressing fossil fuels. Forests are a huge part of the issue." Zero-cut activists such as Hanson, as well as the grassroots activists involved in the Headwaters campaign, have contributed significantly to forest protection in the United States during the last two decades and thus have helped to ensure that these forests continue to sequester carbon from the atmosphere. In addition, Heartwood and other eastern forest protection groups have also turned increasing attention to stopping mountaintop removal, a particularly destructive form of coal mining that is harmful not only for the forests on those mountains, but also in terms of the large volume of greenhouse gases released from burning the resulting coal.

The most notable examples of the role of grassroots biodiversity groups in addressing the climate crisis come from the Center for Biological Diversity. The Center has pioneered the application the Endangered Species Act to global warming, most famously in the case of the polar bear. In the early 2000s, Brendan Cummings began preparing this legal strategy by reviewing the status of species that might be affected by climate change. By 2004, scientific research had revealed that the polar bear was directly imperiled by the effects of global warming. Global warming is most evident in the Arctic where the rising temperature is melting the polar ice cap in the summer to an unprecedented extent, jeopardizing the creatures that live on the polar ice floes. The loss of sea ice had become so pronounced that polar bears, which are excellent swimmers, were drowning because they could not endure the increasingly long journeys between the disappearing ice floes. As Center staff attorney Kassie Siegel summarized, "The Arctic is the Earth's early warning system, warming at twice the rate of the rest of the world. The polar bear is the canary in the coal mine."

In light of these findings, Siegel stepped down from her position fighting urban sprawl in order to write a lengthy petition to list the polar bear as an endangered species under the ESA. Siegel undertook this project even though no foundations were specifically funding the Center's work on this issue at the time.[23] She also encountered concerns from other environmental organizations that the potential controversy around a polar bear listing petition might bolster legislative efforts by opponents of the Endangered Species Act to dismantle that law. However, Siegel saw the polar bear petition as benefiting both the ESA and climate change activism. The plight of the polar bear would help to demonstrate the value of the protections provided by the act. At the same time, the polar bear listing process offered a compelling way to educate the public about the effects of global warming. As Siegel explained,

> Global warming is a complex scientific issue. People perceive it as distant in time and space from them. It's abstract. It's a future threat. And we make it concrete. We make it immediate. People can understand that polar bears are drowning, they're starving, and they're losing their habitat because of global warming.

The Center invited the dominant national environmental organizations working on Arctic protection to sign on as copetitioners in order to bring additional credence to the polar bear listing petition. However, none of these organizations took up the offer at first, so the Center proceeded with filing the petition on its own on February 16, 2005. (Greenpeace and the Natural Resources Defense Council subsequently added their names to the petition in July.) When it became apparent the government was not responding to the petition, the Center and its copetitioners filed suit to compel a decision on listing. In February 2006, they reached a settlement with the government to make a determination by the end of the year. Then on December 27, 2006, Secretary of the Interior Dirk Kempthorne, a noted opponent of environmentalists, announced the results. Although the George W. Bush administration had vigorously resisted most ESA listings, Kempthorne declared that they intended to list the polar bear. It would take another seventeen months, further litigation, and a court order to get that decision finalized, but on May 15, 2008, the federal government officially protected the polar bear as a threatened species under the Endangered Species Act.

One factor in that decision may have been the provision of the ESA that requires that listing decisions be made solely on the best available science. Had the Bush administration denied the listing, it would have been sued by the Center. Kassie Siegel—who became the director of the Center for Biological Diversity's Climate Law Institute—explained, "We were going to litigate the science of global warming in the ESA context and make [legal] precedent on what the science is that has to be used by the federal government." This litigation would have thus jeopardized the Bush administration's efforts to obfuscate the scientific evidence of global warming. Faced with this prospect, the administration chose instead to list the polar bear under the ESA.

The polar bear listing process attracted extensive international media coverage. The Center's listing petition cemented the polar bear's role as a primary icon for the imminent dangers from global warming. The polar bear's iconic status was perhaps best represented by a *Time* magazine issue focused on global warming. The photo on the cover depicted a polar bear on a small ice floe staring plaintively into the surrounding water as blackness closes in ominously from the edges of the image. It was accompanied by the headline, "Be Worried. Be Very Worried."[24] Polar bear imagery also filled the web sites and fundraising materials of the national environmental organizations, including many groups that had opted not to sign on to the Center's listing petition.

Beyond its public education value, the polar bear listing has the potential to directly influence federal policies related to climate change. Under Section 7 of the ESA, the federal government is precluded from engaging in actions that jeopardize endangered species. For example, Siegel explained, "If you are a federal agency and you are approving a coal-fired power plant in Texas, you need to consider the impact of those emissions on polar bears. And I think it is pretty clear that the cumulative impact of the greenhouse gas pollution from coal-fired power plants jeopardizes polar bears." Thus, by challenging permits for new coal-fired plants or through similar applications of the ESA, the enforcement of the polar bear listing can contribute directly to a reduction in greenhouse gas emissions.

While the Center for Biological Diversity pioneered the use of the Endangered Species Act to address global warming, it has not limited its activities to working only with biodiversity protection laws. The Center was a coplaintiff on a case that compelled the government to implement an unenforced provision of the Energy Policy Act of 1992, which sought to reduce U.S. dependence on oil by requiring that 75 percent of the new vehicles purchased each year by federal agencies would have to run on alternative fuels. As a result of that suit, by 2008 there were more than 100,000 alternative fuel vehicles in the federal fleet. The Center also used state-level laws in California to require that new development projects analyze and reduce their impact on global warming. And by end of 2008, the Center was at the forefront of developing powerful new litigation strategies to more fully implement the Clean Air Act in order to regulate greenhouse gas emissions.

Beyond their direct role in curbing greenhouse gas levels, perhaps the most important contribution that grassroots biodiversity groups such as the Center for Biological Diversity offer to the climate campaign may be the role model they provide for other activists. The climate crisis will undoubtedly spark the emergence of a variety of new environmental groups. They would do well to heed the example set by the grassroots biodiversity groups. The climate crisis is an unprecedented threat. Stopping global warming will require large-scale social change. To achieve that level of social change, we will need environmental groups that are bold and unconstrained, that do not fear controversy or conflict, and that advocate for what is ecologically necessary rather than what is narrowly seen as politically realistic. The experiences of the grassroots biodiversity groups during the last twenty years demonstrate that this approach can lead to substantial environmental protection. We should look to the grassroots for bold leadership in addressing the climate crisis as well. These groups embody the essence of environmental champion David Brower's favorite passage from Goethe:

> Boldness has genius, power, and magic in it.

AFTERWORD:

ARRIVAL OF THE

OBAMA ADMINISTRATION

I wrote this book during the second term of the George W. Bush administration and focused on the period between the start of the administration of George H. W. Bush in 1989, through the Clinton presidency of the 1990s, and concluding with the George W. Bush years. However, as this book goes to press in 2009, Barack Obama has been elected president, and Democrats have secured substantial majorities in both houses of Congress. This marks the first time that Democrats have controlled all three arenas since the first two years of the Clinton administration. The experiences of the environmental movement during the Clinton era may thus offer some particularly useful lessons to inform activism during the Obama administration. If that period is any guide, then we are likely to see a greater distinction between the grassroots groups and the national environmental organizations reemerge in the coming years. Indeed, there has been evidence of that trend developing even before Obama took office.

In December 2008, the president-elect announced his intention to select Senator Ken Salazar (D-CO) as secretary of the Interior. Other candidates under consideration for that position had much clearer biodiversity protection credentials, whereas when Salazar was Colorado's attorney general, he had threatened to sue to the U.S. Fish and Wildlife Service if the agency protected the black-tailed prairie dog under the Endangered Species Act. The Center for Biological Diversity responded to Obama's announcement with a press release criticizing the choice. The Center's executive director Kieran Suckling went on record saying, "[The Obama] team will be weakened by the addition of Ken Salazar, who has fought against federal action on global warming, against higher fuel efficiency standards, and for increased oil drilling and oil subsidies."[1] Suckling's critique attracted widespread media coverage, but often those news articles included counterpoints from national environmental organizations praising the selection of Salazar. Defenders of Wildlife, Environmental Defense Fund, National Audubon Society, National Parks Conservation Association, Natural Resources Defense Council, National Wildlife Federation, Sierra Club, and The Wilderness Society all issued statements supporting Salazar. One commentator noted, "National Audubon explained its support for Salazar as motivated by a desire to preserve 'access' to the Secretary."[2] In contrast, WildEarth Guardians (formerly Forest Guardians) organized a sign-on letter to Obama from grassroots biodiversity groups and scientists highlighting their concerns regarding Salazar's problematic record on endangered species protection.

Once again, the distinction between the insider approach of the national environmental organizations and the outsider approach of the grassroots groups appears likely to define their respective approaches to environmental advocacy during the Obama presidency. Barack Obama ran for president on a message of hope. Based on the recent history of biodiversity activism, the best hope for substantial environmental protection amid the Obama years depends on continued bold advocacy from the grassroots groups.

APPENDIX
ORIGINS OF FOUR
BIODIVERSITY PROTECTION LAWS

Following the first Earth Day in 1970, there was a burst of federal legislation on environmental protection. Twenty-three major environmental laws were enacted during the decade of the 1970s, including the National Environmental Policy Act, the Marine Mammal Protection Act, the Endangered Species Act, and the National Forest Management Act. These four laws in particular would be used extensively in subsequent decades by grassroots biodiversity groups as the basis for litigation. This environmental legislation is often presented as one of the foremost accomplishments of the national environmental organizations and as a reflection of their political clout following Earth Day. However, closer examination of the legislative history reveals that the dominant national environmental organizations played relatively minor (and at times even counterproductive) roles in the passage of these laws.

National Environmental Policy Act
The National Environmental Policy Act slightly predated the first Earth Day, having been signed by President Nixon in January 1970. It received little debate in Congress and passed by overwhelming majorities in both the House and Senate. This is arguably the most wide-reaching piece of federal environmental legislation. In addition to creating the Council of Environmental Quality within the executive branch, it established the requirement of an environmental impact statement for every major federal project. As part of the environmental impact statement, federal agencies must assess the environmental consequences of their proposed actions and consider a range of alternatives. This process provides substantially increased opportunities for public participation in agency policy making, both by creating a detailed public record of environmental impacts and also by providing for increased public input and oversight. Thus the National Environmental Protection Act became the foundation for much of the environmental litigation against federal agencies. Environmental scholar Robert Gottlieb described this legislation as "the cornerstone of contemporary environmental policies," but noted that it was enacted "with little input from any of the mainstream [environmental] groups who eventually constituted the Group of Ten."[1]

Marine Mammal Protection Act

The Marine Mammal Protection Act became law in the 1972. It reflected a high level of public interest in the welfare of whales and dolphins. One key impetus for the Marine Mammal Protection Act was the revelation that hundreds of thousands of dolphins were being killed each year by the U.S. tuna fishing fleet in the Eastern Tropical Pacific. There the fishers would deliberately encircle dolphins in nets because large schools of tuna were often found swimming with the dolphins. Although the fishery did not seek to harvest the dolphins, many were killed when they became entangled in the nets and drowned. It was estimated that about 5 million dolphins were killed this way in the late 1950s and 1960s.[2] When information about the toll on dolphins was published in the late 1960s, there was a substantial public outcry. During the early 1970s, Congress reported receiving more mail about marine mammal protection than any other issue other than the Vietnam War.[3]

In response, a bill called the Ocean Mammal Protection Act was introduced in Congress in 1971. The prospective legislation was notable for its comprehensive ban on the killing of marine mammals and the importation of marine mammal products. At first the bill attracted significant political momentum, but then it ran into opposition from a coalition of conservation groups representing wildlife managers, including the National Audubon Society and the World Wildlife Fund. Wildlife managers objected to the comprehensive moratorium, which was seen as divergent from the "maximum sustained yield" approach that had dominated wildlife management up until that time. The coalition succeeded in derailing the Ocean Mammal Protection Act and replacing it with a less restrictive bill called the Ocean Mammal Management Act, which sought to manage marine mammals within a sustained yield approach, while exempting the tuna fleet from these regulations. This bill was poised to be passed by the House of Representatives in December 1971 when a last minute flurry of media outreach by Friends of Animals—a small animal rights organization with one person on staff in DC—led to *Washington Post* and *New York Times* editorials against the Ocean Mammal Management Act. The bill was subsequently defeated.[4]

Then in 1972, the Marine Mammal Protection Act was introduced as a compromise bill, combining elements of both the Ocean Mammal Protection Act and the Ocean Mammal Management Act. It became law by the end of the year. The Marine Mammal Protection Act returned to a moratorium approach, but included various exemptions, including a provision allowing for a long phase-out period for the dolphin bycatch of the tuna fleet. As a result, tens of thousands of dolphins continued to be killed by U.S. fishers each year through the 1970s and 1980s. Thus in the case of the Marine Mammal Protection Act, two of the national environmental organizations (National Audubon Society and World Wildlife Fund) played a counterproductive role that led to a weakening of biodiversity protection compared to the original legislation. Moreover, the much more problematic legislation originally promoted by those two organizations was defeated by a small animal rights group, Friends of Animals, rather than any of the established national environmental organizations.

Endangered Species Act

The Endangered Species Act of 1973 is the most powerful biodiversity protection law in the United States. Indeed it has been described as "the most far-reaching wildlife statute ever adopted by any nation."[5] The 1973 Endangered Species Act was considerably more powerful than two prior endangered species laws enacted in 1966 and 1969. It not only precluded the killing of species at risk of extinction but also protected their habitat and included significant restrictions on any federal project that might harm endangered species. The impetus for this new legislation can be traced to President Nixon's call for "a stronger law to protect endangered species of wildlife."[6] This call appears to have emerged out of the Nixon administration's frustrated efforts to get whales protected under the Endangered Species Conservation Act of 1969.[7] The new legislation faced no notable opposition in Congress or from industry. No commercial interests testified against the bill as it was proceeding through congressional hearings. It moved quickly through Congress and was passed with unanimous support in the Senate and only twelve votes against it in the House of Representatives.

Yet unlike the much earlier case of the Wilderness Act where the role of groups such as the Sierra Club and The Wilderness Society was clearly evident (see chapter 2), one is hard pressed to find evidence of the national environmental organizations playing a significant role in the passage of the Endangered Species Act. Indeed, in policy analyst Steven Yaffee's account of the legislative history, the groups that he listed as being directly involved in lobbying for a strong endangered species law were animal rights groups and relatively minor environmental groups—the Fund for Animals, the Society for Animal Protective Legislation, the National Parks and Conservation Association, and the Committee for the Preservation of Tule Elk[8]—rather than the dominant national environmental organizations. (For a more detailed discussion of the provisions of the Endangered Species Act and a history of the subsequent amendments, see chapter 5.)

National Forest Management Act

The National Forest Management Act of 1976 provided increased opportunities for public oversight of the Forest Service. It would become a key tool of forest protection activists and ultimately was the legal cornerstone for Judge William Dwyer's sweeping injunctions against logging on national forests in the Pacific Northwest in 1989 and 1991 (see chapter 4). However, those later successes stood in contrast to the National Forest Management Act's troubled origins. Unlike with the Endangered Species Act, the involvement of national environmental organizations was readily apparent in the legislative history of the National Forest Management Act, but the effect of their involvement was problematic.

The National Forest Management Act's origins can be traced to an early grassroots campaign against clearcutting on national forests. Clear cutting entails removing all of the trees within an area and thus is the most severe form of deforestation. Originally the Forest Service had criticized the timber industry's use of clearcutting, but as logging levels rose on national forests in the decades following World War II, the Forest Service increasingly adopted extensive use of clearcuts as well. This provoked growing criticism of the Forest Service. As historian Paul Hirt recounted, "By 1970, grassroots revolts against the Forest Service's timber program had erupted from New Hampshire to California, from Alaska to Georgia. Local efforts, as usual, presaged national campaigns by the organized environmental groups."[9] Then in 1973, the West Virginia division of the Izaak Walton League filed suit against the Forest Service for permitting clearcutting in the Monongahela National Forest, charging that clearcutting violated the provisions of the Forest Service's founding Organic Act of 1897. The courts ruled in favor of the plaintiffs. This ruling held the potential of leading to a shut down in clearcutting throughout the national forest system. Anti-clearcutting litigation then spread to other parts of the country, but it faced resistance from the national environmental organizations. As Hirt explained, "Grassroots organizations, such as the Texas Committee on Natural Resources headed by environmental lawyer Ned Fritz, pursued the Izaak Walton League's successful strategy only to find themselves opposed by the national groups who feared that additional favorable rulings would strengthen the hand of those mobilizing to repeal the 1897 act. The prime strategy of the Sierra Club was to retain the old law by any means possible."[10]

As the advocates of clearcutting promoted legislation to repeal or revise the Organic Act, the Sierra Club's status quo approach became increasingly untenable. The Club and other national organizations decided to throw their support behind a forestry reform bill offered by Senator Jennings Randolph (D-WV). However, this bill languished in the face of weaker legislation proposed by Senator Hubert Humphrey (D-MI). The Humphrey bill offered some environmental protection measures, but Hirt noted that it was "primarily motivated by the need to legalize clearcutting so that Forest Service could get back to its task of rapidly liquidating its native forests and substituting even-aged forests of commercial species."[11] The Izaak Walton League, National Audubon Society, Sierra Club, and The Wilderness Society opposed the Humphrey bill, while the more conservative National Wildlife Federation testified on its behalf.[12] Yet the Humphrey bill was passed into law as the National Forest Management Act of 1976 over the objections of many of the nationals. Thus, while this legislation would go on to play an important role in grassroots forest protection activism, it is difficult to see the passage of this bill as an example of the political influence of the national environmental organizations.

CHAPTER NOTES

Chapter 1

1. For an account of Sea Shepherd's 1990 driftnetting campaign, see Watson, "Tora! Tora! Tora!"
2. Kiefer, "Owl See You in Court," 22.
3. Center for Biological Diversity, *Celebrating Fifteen Years of Protecting the Wild*, 1.
4. Takacs, *The Idea of Biodiversity*.
5. Daily, ed., *Nature's Services*.
6. Devall and Sessions, *Deep Ecology*.
7. Broswimmer, *Ecocide*.
8. Hirt, *A Conspiracy of Optimism*.
9. See Dowie, *Losing Ground*; Gottlieb, *Forcing the Spring*.
10. Dowie, *Losing Ground*, 6. Writing in the early 1990s, Mark Dowie highlighted the grassroots environmental justice activists as the principal alternative to the national environmental organizations. Environmental justice groups focused primarily on the disproportionate impact of toxic waste and other pollution on low-income communities and communities of color. Grassroots environmental justice activists rightly attracted significant attention from the media and scholars. However, the environmental justice groups were not the only form of grassroots activism in the environmental movement. While they were only just starting to have an impact at the time of Dowie's research, the grassroots biodiversity groups ultimately demonstrated that there could also be an effective grassroots approach to the protection of forests and wildlife.
11. The exclamation point was specifically added to the name of the group by its founders in order to convey their vehemence. It also reflected their cavalier attitude toward rules, including the rules of grammar. The exclamation point thus does not serve a punctuation role in sentences in which the name "Earth First!" is used.
12. Although the term is sometimes used to disparage social change activists, the word "radical" is derived from the Latin word for root ("radix") and in a social-change context it originally connoted going to the root of a problem. This definition raises a key question—why wouldn't all social change groups strive to be radical? If a group is not seeking to address the root of a problem, then it must somehow be constrained from doing so. It is therefore useful to examine the ways that social movements are constrained in their advocacy and under what circumstances they can avoid at least some of those constraints.
13. "Mainstream" is often used as the antonym for radical. However, that term is a bit misleading in this context because it suggests that the radical position lacked widespread support, which was often not the case for biodiversity issues. The full protection of imperiled species and forests enjoyed broad public support. Indeed, some of the positions of the so-called radical activists had higher levels of public approval than the less protective measures endorsed by the national environmental

organizations (see, for example, the polling results for zero cut and the Roadless Rule in chapter 4). In place of mainstream, the term "moderate" offers a counterpoint to radical that serves as a more accurate description of the organizations at the more constrained end of the spectrum. When used as verb rather than an adjective, moderate has a similar meaning to "constrain," that is, to hold back.

14. Aleshire, "A Bare-Knuckled Trio," 9.

15. Outsider groups sometimes work on political change as well, but they approach it quite differently from insider groups. Rather than hoping that a politician will promote legislation as a favor in exchange for the group's support, outsider groups compel politicians to act on an issue by creating situations that the politicians simply cannot ignore. For example, as described in chapter 3, when more than 1,000 people demonstrated that they were willing to get arrested to block the logging of Headwaters Forest, the federal government was compelled to intervene to protect the forest. Moreover, when outsider groups engage in outreach to politicians, they generally do so through mobilized constituents rather than professional lobbyists. Since they do not depend on maintaining privileged access to politicians, outsider groups are likely to promote legislation that offers stronger environmental protections than the more compromise-based measures endorsed by insiders, as evident with the forest and endangered species protection legislation described in chapters 4 and 5.

16. To be clear, these descriptions of the grassroots and the nationals are ideal types. There are of course some exceptions and variations to these characterizations. For example, there were some small groups that attempted to use an insider strategy, such as the Rose Foundation (see chapter 3). Nonetheless, an outsider approach was a defining characteristic for the primary grassroots biodiversity groups in this study. Likewise, not all national environmental organizations behaved in exactly the same manner. For example, the Environmental Defense Fund was seen as particularly conservative and willing to broker deals with antienvironmental corporations and politicians, whereas on the other end of the spectrum, Defenders of Wildlife was sometimes willing to be a coplaintiff in litigation by grassroots biodiversity groups (see chapter 5). Nonetheless, the actions of all of the national organizations were circumscribed by their reliance on an insider strategy. In the case of Defenders, its insider approach ultimately led it to hire a former Clinton administration official as vice president of the organization in the early 2000s, and Defenders' endangered species policies subsequently became more aligned with Environmental Defense Fund (see chapter 5).

17. See Michels, *Political Parties*; Piven and Cloward, *Poor People's Movements*.

18. For examples, see Dowie, *American Foundations*; Faber and McCarthy, eds., *Foundations for Social Change*; and INCITE! Women of Color Against Violence, eds., *The Revolution Will Not Be Funded*.

19. I address the idea of "biocentrism" and its counterpart "anthropocentrism" in the context of examining Earth First!'s movement culture in chapter 2. As I discuss in that chapter, biocentrism is the belief that other species of wildlife have an inherent right to exist, whereas an anthropocentric perspective values other species only in terms of their usefulness to humans. Some scholars have put a strong emphasis on the role of these ideologies in the behavior of environmental groups, contending that the biocentrism/anthropocentrism dichotomy is central to distinction between Earth First! and associated grassroots groups on the one hand, and the national environmental organizations on the other hand (see, for example, Lee, *Earth First!*). However, based on my case studies, I do not adopt this emphasis on ideological distinctions and in fact believe that it can obfuscate

the underlying dynamics. I observed that many environmental activists in both radical grassroots groups and moderate national organization believe in protecting biodiversity for its own sake, rather than solely because of its usefulness to humans. However, the staff of the nationals were more likely to publicly articulate their goals in anthropocentric terms because they believed that was necessary to maintain their access to politicians and policy makers as part of these organizations' use of an insider strategy. Because of this strategy, they were also more likely to limit their stated biodiversity protection goals to what they saw as politically realistic. By comparison, grassroots groups that were not dependent on an insider strategy were freer to endorse stronger goals that were not politically expedient, and to express their support for these goals in biocentric terms. In this regard, the distinction between the positions endorsed by the national organizations and the grassroots groups has less to do with whether they have an anthropocentric or biocentric ideology than with whether they think that advocating a particular position is compatible with their underlying strategy for achieving biodiversity protection. In short, what is often presented as an ideological divide is instead the reflection of the differences between an insider and outsider strategy.

20. With regard to Congress, there were also nearly equal periods of dominance by each political party during this time. Democrats controlled both houses of Congress from 1989 through 1994. Republicans controlled both houses of Congress from 1995 through 2000, and again from 2003 through 2004. In 2001 through 2002, Democrats controlled the Senate while Republicans controlled the House of Representatives. Also, during the time frame covered by this study, each party experienced a two-year period in which it controlled both houses of Congress and the White House simultaneously. For Democrats, this was from 1993 through 1994. For Republicans, it was from 2003 through 2004. That said, for the purposes of this study, I emphasize the party affiliation of the president because of the primary role of the executive branch in implementing environmental laws.

21. Shellenberger and Nordhaus, "The Death of Environmentalism."

22. Pickett, "Is Environmentalism Dead?"; Nordhaus, Shellenberger, and Werbach's omission of these accomplishments was all the more egregious because all three authors had been directly involved with grassroots biodiversity campaigns: Nordhaus formed the Headwaters Sanctuary Project in the latter part of the Headwaters campaign; Shellenberger was a professional consultant employed by both the Headwaters campaign and the zero-cut campaign; and Werbach was a member of the Sierra Club board of directors during the John Muir Sierrans' efforts to transform the Club's position on zero cut. Moreover, as we will see in chapters 3 and 4, each of them played problematic roles in these campaigns.

Chapter 2

1. The membership of the Group of Ten left out several other large environmental organizations, including the Nature Conservancy, World Wildlife Fund, Defenders of Wildlife, and Greenpeace. The Nature Conservancy formed in 1951 and primarily protects land through direct acquisition rather than advocacy. The World Wildlife Fund was founded in Switzerland in 1961 and focuses on international conservation projects. While it has a large membership in the United States, it plays a relatively small role in U.S. environmental policy. Defenders of Wildlife was deliberately excluded from the Group of Ten by the director of the National Wildlife Federation, who opposed Defenders' antihunting position.

Begun in 1947 under the name Defenders of Fur Bearers, it was the last major conservation-era wildlife protection group to form. Defenders was still comparatively small in the 1980s but was well integrated into environmental policy circles in Washington, DC by the 1990s. In the 1990s, the Group of Ten was renamed the "Green Group" and expanded to include these and other large national environmental organizations. The largest environmental organization not included in either the Group of Ten or the Green Group was Greenpeace, which began as an antinuclear group in 1971 but it soon became better known for its work to protect whales and other marine mammals. By the 1980s, its worldwide membership was much larger than any other environmental group. Greenpeace's use of civil disobedience set it apart from more conventional, large environmental organizations. At the same time, its extensive organizational bureaucracy distinguished it from Earth First!. These characteristics made Greenpeace an anomaly among environmental organizations. It thus does not readily fit into the any of the three categories of biodiversity activism profiled in this chapter.

2. For a discussion of the motivations of biologists involved in biodiversity protection, see Takacs, *The Idea of Biodiversity*.
3. With ideal types, there are inevitably some specific exceptions to these broad descriptions.
4. Fox, *The American Conservation Movement*, 214; 279.
5. Cohen, *The History of the Sierra Club*, 178.
6. Fox, *The American Conservation Movement*, 288.
7. Hirt, *A Conspiracy of Optimism*, 231.
8. Gottlieb, *Forcing the Spring*, 107.
9. Ibid., 106.
10. Brulle, *Agency, Democracy, and Nature*, 186.
11. Gottlieb, *Forcing the Spring*, 137.
12. Mitchell, "From Conservation to Environmental Movement," 97.
13. Ibid., 99.
14. Quoted in Dowie, *Losing Ground*, 76.
15. Mitchell, Mertig, and Dunlap, "Twenty Years of Environmental Mobilization," 20.
16. Mitchell, "From Conservation to Environmental Movement," 104.
17. Foreman, *Confessions of an Eco-Warrior*, 13–14.
18. Dowie, *Losing Ground*, 44.
19. Ibid., 43.
20. Ibid., 44.
21. See Brulle, *Agency, Democracy, and Nature* for a more detailed analysis of the impact of direct mail fundraising on environmental organizations. See also Dowie, *Losing Ground*.
22. Dowie, *Losing Ground*, 45.
23. See Gottlieb, *Forcing the Spring*.
24. Fox, *The American Conservation Movement*, 334.
25. The Sierra Club was a partial exception to this characterization, as discussed in chapter 4.
26. See Piven and Cloward, *Poor People's Movements*.
27. Mitchell, Mertig, and Dunlap, "Twenty Years of Environmental Mobilization," 15.
28. Ibid., 13.
29. The one exception was Mike Roselle, whose background was primarily with counterculture groups such as the Yippies. Roselle would become a prominent figure in the changes that occurred in Earth First! in the late 1980s and 1990s as

the other cofounders stepped back. Roselle also served as a mentor to some of the new grassroots biodiversity groups that formed during that time. And he became an informal adviser to some of the key funders for those groups, such as Turner Foundation Director Peter Bahouth.

30. Scarce, *Eco-Warriors*, 23.
31. See Foreman, *Confessions of an Eco-Warrior*.
32. Lee, *Earth First!*, 29.
33. The results of the RARE II were subsequently overturned in court when Huey Johnson, California's secretary of resources, successfully sued under the National Environmental Protection Act to block the Forest Service's plans to develop roadless areas in California. Johnson pursued this litigation even though The Wilderness Society and Natural Resources Defense Council tried to talk Johnson out of filing the lawsuit. See Zakin, *Coyotes and Town Dogs*.
34. Quoted in Scarce, *Eco-Warriors*, 24.
35. Zakin, *Coyotes and Town Dogs*, 311.
36. Ingalsbee, "Earth First! Consciousness in Action," 151–52.
37. Quoted in Zakin, *Coyotes and Town Dogs*, 307.
38. For an early description of this phenomena in relation to the women's movement, see Freeman, "The Tyranny of Structurelessness."
39. See Devall and Session, *Deep Ecology*.
40. See Takacs, *The Idea of Biodiversity*.
41. Lee, *Earth First!*, 116.
42. Ibid., 85.
43. Ibid., 116.
44. Mitchell, Mertig, and Dunlap, "Twenty Years of Environmental Mobilization," 118.
45. In some regards, their approach to outsider strategy went even further than that of the founders of Earth First!. For Dave Foreman, Earth First!'s role was to shift the terms of the environmental debate toward greater protection and allow the moderate national organizations to look reasonable while asking for more (see Foreman, *Confessions of an Eco-Warrior*). Thus, his approach ultimately still depended on the nationals to attain the environmental protections through their political deal-making. However, subsequent generations of Earth First!ers sought to achieve biodiversity protection directly without depending on nationals. While some activists pursued this goal within Earth First! in the 1990s (see chapter 3), others formed grassroots biodiversity groups to do so.
46. See Dowie, *American Foundations*.
47. Gottlieb, *Forcing the Spring*.

Chapter 3

1. Harris, *The Last Stand*, 142.
2. Champion, "Redwood Cutting Plan Provokes a Protest," A24.
3. Walters, "California's Chain-Saw Massacre."
4. Quoted in Skow, "Redwoods: The Last Stand," 59.
5. The one known case of an injury from a tree-spike was a Louisiana-Pacific mill-worker named George Alexander, who was seriously injured in 1987. At first, the Louisiana-Pacific timber corporation widely publicized Alexander's plight in order to condemn Earth First!, but it soon became apparent that Earth First! was most likely not responsible for the spike. The tree was very small, not the old-growth trees that Earth First! sought to protect. Moreover, from the height of the spike it was apparent that the tree had been spiked after it had been cut. After Alexander refused to go on a publicity tour to denounce Earth First!, he had to

sue Louisiana-Pacific to get compensation for his injuries. He was subsequently transferred to the night shift at the mill and was later laid off when the mill closed down. See Bari, *Timber Wars*.
6. Quoted in Bari, *Timber Wars*, 310.
7. Ibid., 311.
8. Ibid., 224.
9. Ibid.
10. Ibid., 78.
11. Environmental Protection Information Center, "Summary of EPIC Forestry Lawsuits 1982 to 1996."
12. Ball, "State Forestry Update."
13. Pickett, "Of Ballot Initiatives, Political Deals, Millionaires and Ancient Trees."
14. Bailey, "Timber Accord: What's the Big Deal?," B5.
15. Maxxam had previously illegally logged the grove over Father's Day weekend of 1992, after California had listed the murrelet under the state-level California Endangered Species Act and EPIC had successfully challenged Maxxam's Owl Creek timber harvest plan in state court. EPIC got the Father's Day logging stopped after three days, but the company made $1 million from that timber and did not face any serious penalties.
16. Ghent, "Thanksgiving Massacres on West Coast," 1.
17. Skow, "Redwoods: The Last Stand."
18. While both the Rose Foundation and As You Sow Foundation featured "foundation" in their names, these were not traditional foundations based around a large endowment. Instead they would redistribute funds from environmental litigation settlements and private donors. And unlike most foundations, the staff of these organizations also worked directly on activist projects.
19. Knickerbocker, "Debt-for-Nature Swap Proposed between Uncle Sam and Loggers," 1.
20. Roemer, "Attorney to the Tree," 7.
21. Ghent, "Don't Take Orders from Anyone."
22. Cherney, "If It Ain't Broke, Don't Fix It."
23. Ghent, "Mass Action for Headwaters," 28.
24. Bundy, "The People's History of Activism from Headwaters."
25. Karen Pickett interview.
26. Barnum, "Protests Widen Over Headwaters,"A15.
27. Spring and Arcky, "Owl Creek Rises," 15.
28. Quoted in Barnum and Diringer, "Juggling Trees, Wildlife and Property Rights," SC1.
29. Quoted in Bevington, "Lessons from the HFCC in Campaign Coalition Organization," 3.
30. Ibid., 2.
31. Nordhaus's future collaborator on that essay, Michael Shellenberger, had been brought into the Headwaters campaign by Tom Van Dyck to serve as a professional media consultant.
32. The problems of the Headwaters campaign in this regard reflected a pattern in other American social movements identified by sociologist Francesca Polletta in *Freedom Is an Endless Meeting*. According to Polletta, the particular form of participatory democracy in a social movement organization reflects the form of relationship between the founding members of that group. However, when the group grows rapidly and diversifies, that relationship may break down as a basis for group decision making.

33. Quoted in Kay, "Fight for Old Growth Hasn't Ended."
34. *Chronicle* Staff Report, "Headwaters Protest Turns Ugly."
35. Some grassroots activists were involved in outreach to elected officials, but they took a notably different position from the insider groups. For example, Cecilia Lanman of EPIC was a frequent participant in meetings with federal officials, yet she remained an outspoken critic of the deal. Likewise Darryl Cherney of Earth First! went on lobbying trips to DC, though he emphasized that the grassroots activists took a fundamentally different approach to political outreach than that of professional environmental lobbyists such as the Sierra Club staff.
36. Clifford, "Deal to Save the Headwaters Could Fail," A6.
37. Barnum, "Toppled Tree Kills Logging Protester in Humboldt," A21.
38. *Earth First! Journal*, "If Words Could Kill . . . ," 19.
39. Pickett, "Headwaters Activist Killed," 1.
40. *Earth First! Journal*, "What Happened That Day: Accounts from Earth First!," 19.
41. In other parts of the country, frustration with the civil disobedience approach led a fraction of activists to form the Earth Liberation Front, which engaged in clandestine property damage to businesses involved in environmentally destructive activities. Their actions attracted extensive media coverage, but ultimately many of the activists were arrested and jailed during the George W. Bush administration. See Rosebraugh, *Burning Rage of a Dying Planet*.
42. Jakubal, "Green Steel."
43. Pickett, "Redwood and Steel," 41.
44. Wreckin' Ball, "Union/Eco Alliance Costs Hurwitz Half a Mil."
45. Thompson, "Baked Goods."
46. Apple, ed., *Pie Any Means Necessary*.
47. Martin, Lochhead, and Gledhill, "Headwaters Deal Lauded as Model Agreement," A1.
48. Martin and Curiel, "Last-Minute Headwaters Deal OKd," A1.
49. Martin, Lochhead, and Gledhill, "Headwaters Deal Lauded as Model Agreement."
50. Salladay, "Biologists Butted in to Save Redwoods."
51. Bailey, "Why I Should Feel Good (Even If I Don't)."
52. California Public Employees for Environmental Responsibility, *California's Failed Forest Policy*, 10.
53. *Examiner* Editorial Board, "Salvation Is at Hand."
54. Hill, *The Legacy of Luna*, 106.
55. During Thanksgiving 2000, a vandal cut through about 60 percent of Luna's trunk with a chainsaw. The damage reduced the nutrient flow within the tree and left Luna vulnerable to being blown over by the wind. Arborists developed a sophisticated set of braces and supports to prevent blow over. The cut will reduce the tree's lifespan, though Luna is still standing at the time of this writing and may continue to do so for years or even centuries to come. See Salter, "Attack on Luna Another Test for Hill."
56. Lee, "State Legislature Debates While Treesitters Hold Strong," 8.
57. Pickett, "Labor and Environmentalists," 5.
58. Goodman, "Standing Up for Democracy."
59. Kegley, "Victory! Victory! Victory!"
60. Taylor, "Corporate Irresponsibility United Us," 10.
61. Environmental Protection Information Center, "EPIC's Historic Case Goes to Trial."
62. Lifsher, "Lawyer Goes against the Grain in Logging Fight with Maxxam."

63. Beach, *A Good Forest for Dying*.
64. Bari, "FBI Bomb School and Other Atrocities."
65. See Bari, *Timber Wars*.
66. Salladay, "Tree Hugger's Hunger Strike at 40 Days," A2.
67. Gunnison and Lucas, "Critics Say Davis Kowtows to Donors."
68. Quoted in California Public Employees for Environmental Responsibility, *California's Failed Forest Policy*, 12.
69. Abate, "Hurwitz Wins FDIC Lawsuit."
70. Lucas, "Timber Report Called Faulty."
71. Martin, "D.A. Hits Timber Firm with Fraud Charges," A1.
72. Gurnon, "One Year Later."
73. Sims, "Case Dismissed."
74. Mark Lovelace interview.
75. Driscoll, "State Water Board: PL's Peril Its Own Making."
76. Abate, "Lumber Dispute Coming to a Head," E1.
77. The counterculture values of that community also helped to foster a social context in which women could take on many of the leadership positions within the Headwaters campaign.
78. Arquilla and Ronfeldt, eds., *Networks and Netwars*.
79. It should also be noted that Headwaters activists came close to more fully realizing their goals through the Forests Forever initiative at a time when there were Republicans in both the White House and the California governor's mansion.

Chapter 4

1. See Hirt, *A Conspiracy of Optimism*.
2. See Hanson, *Ending Logging on National Forests: The Facts*.
3. For more detailed accounts of this period, see Durbin, *Tree Huggers* and Yaffee, *The Wisdom of the Spotted Owl*.
4. Zakin, *Coyotes and Town Dogs*, 261–62.
5. Ibid., 233.
6. Dubin, *Tree Huggers*, 38.
7. Zakin, *Coyotes and Town Dogs*, 236.
8. Ibid., 238.
9. Ibid., 262.
10. Ibid., 236.
11. Ibid., 311.
12. At that meeting, Talberth also met his future wife, Charlotte Levinson. Levinson (later Charlotte Talberth) was an activist in the Cathedral Forest Action Group and she was also connected with a small but influential family foundation. In the mid-1980s, she became a director of the Levinson Foundation. (She also served on the board of directors of ONRC during the late 1980s and early 1990s.) During the 1990s, Charlotte Talberth became one of the earliest and most outspoken advocates for zero cut within the environmental grantmaking community.
13. John Talberth interview.
14. Ibid.
15. Durbin, *Tree Huggers*, 89.
16. A board foot is a standard measure for timber that connotes the volume of a board of wood one foot long, one foot wide, and one inch thick.
17. U.S. Forest Service, "Cut and Sold Reports."
18. Durbin, *Tree Huggers*, 182–83.

19. Yaffee, *The Wisdom of the Spotted Owl.*
20. Quoted in Yaffee, *The Wisdom of the Spotted Owl*, 108.
21. Durbin, *Tree Huggers.*
22. Yaffee, *The Wisdom of the Spotted Owl*, 108.
23. Quoted in Durbin, *Tree Huggers*, 91.
24. Yaffee, *The Wisdom of the Spotted Owl*, 116.
25. Zakin, *Coyotes and Town Dogs*, 267.
26. Durbin, *Tree Huggers*, 107.
27. Ibid., 108.
28. Ibid., 110.
29. *Seattle Audubon Society v. Evans*, 1090.
30. Yaffee, *The Wisdom of the Spotted Owl.*
31. Durbin, *Tree Huggers*, 207–8.
32. Yaffee, *The Wisdom of the Spotted Owl.*
33. Quoted in Durbin, *Tree Huggers*, 211.
34. Ibid., 210.
35. Ibid., 211.
36. Ibid.
37. Ibid., 214.
38. For a more detailed history of Protect Our Woods, see Higgs, *Eternal Vigilance.* An immediate predecessor to, and inspiration for, POW was a grassroots group in Bloomington, Indiana, known simply as Forest Watch, whose cofounders included Denise Joines and Jeffrey St. Clair. Joines later became a program officer for the Wilburforce Foundation, an important funder for grassroots biodiversity groups in North America. St. Clair became an outspoken critic of the national environmental organizations. His writings on the subject were collected in *Been Brown So Long It Looked Like Green to Me* and *Born under a Bad Sky.*
39. John Muir was the earliest noted proponent of this position, and many Earth First!ers advocated for this goal as well, but Protect Our Woods was the first organization to formally endorse it.
40. Orin Langelle, "Shawnee Showdown!"
41. Andy Mahler interview.
42. Like many other grassroots biodiversity groups with limited resources, Heartwood often relied on lawyers willing to represent them without charge (pro bono). When such lawyers could not be found, Heartwood activists without law degrees would sometimes pursue their cases without a lawyer (pro se).
43. Tim Hermach interview.
44. Ibid.
45. This David Orr should not to be confused with David W. Orr, an author and professor of environmental studies at Oberlin College.
46. See Bader, "Northern Rockies Ecosystem vs. Washington, DC Political System." The Sierra Club became embroiled in further controversy when instead it initially endorsed an alternate bill that not only protected far less wilderness than the Northern Rockies Ecosystem Protection Act proposal, but it also included "release" language that actively precluded 4.5 million acres of roadless areas from being later designated as wilderness. Margaret Hays Young reported that the Sierra Club lobbyists argued that support for this less protective bill was necessary to get the congressman who had introduced it reelected, despite that congressman's weak environmental voting record. See Young, "Wilderness Now!"
47. Margaret Hays Young interview.
48. The name deliberately paralleled the recently formed Association of Forest

Service Employees for Environmental Ethics, a support network for whistle-blowers within the Forest Service.

49. Orr, "Reforming the Sierra Club."

50. Young, "Nightmare on Polk Street."

51. Quoted in Dowie, *Losing Ground*, 218–19.

52. Rozek, "But I'm Just One Person, What Can I Do?"

53. Tahoma, "Cascadia Rising."

54. Brock Evans interview.

55. Quoted in Brian Tokar, *Earth for Sale*, 97.

56. Chad Hanson's mother and sister—Connie and Amy—also founded a network of Christian environmental activists called Christians Caring for Creation, which helped to mobilize religious groups to endorse the complete protection of national forests from logging. As a result of their outreach, in 1999, the National Council of Churches—representing 104 denominations and 50 million people—approved a resolution calling for an end to commercial logging on national forests.

57. The Sierra Club never embraced the phrase "zero cut." Club staff contended that the phrase was inaccurate because most zero-cut advocates did not oppose some forms of noncommercial cutting of brush and small trees to repair ecological damage caused by previous logging and fire suppression. Therefore, the Sierra Club staff chose the cumbersome but technically more accurate phrase, "end commercial logging." This phrase was subsequently adopted by many zero-cut activists. However, for the sake of continuity, I will continue to use zero cut to describe this position through the remainder of this chapter.

58. Some other notable groups that began as Earth Island Institute projects include the Bluewater Network, Rainforest Action Network, Sea Turtle Restoration Project, and Urban Habitat.

59. Hanson, *Ending Logging on National Forests: The Facts in the Year 2000*, 1.

60. Hanson, *Ending Logging on National Forests: The Facts*, 1.

61. Chad Hanson interview.

62. Rene Voss recalled that the meeting took place on the exact date on which, ninety-nine years earlier, President McKinley had signed into law the appropriations rider that first permitted logging to occur on the national forests (at the time called "forest reserves"), which had previously been protected from all logging.

63. For example, paper can be made from the agricultural waste left over after harvesting fields. (This material is usually burned, which can cause severe air pollution in farming communities.) However, ag-waste paper is not currently economically competitive with wood paper, in part because the timber industry has benefited from decades of indirect government subsidy through the Forest Service timber sales program. The Environmental Protection Agency fund that would be established by NFPRA was intended to help level the playing field for tree-free alternatives.

64. For an analysis of an earlier effort to elect reform candidates to the Sierra Club board of directors, see Devall, "The Governing of a Voluntary Organization."

65. Brower, *For Earth's Sake*, 436.

66. During these two years, two other candidates were elected who had the trappings of reformers but did not run as JMS petition candidates—Adam Werbach and Dave Foreman. Both proved to be disappointments to the John Muir Sierrans. Werbach, who later became an outspoken proponent of the "Death of Environmentalism" essay, attracted publicity as the youngest Sierra Club board president, but the John Muir Sierrans described him as, at best, an unreliable ally. Former Earth First!er Dave Foreman confounded the John Muir Sierrans by

becoming a leading opponent of their zero-cut ballot initiative in 1996.

67. Michael Dorsey had initially been approached by the Club's nominating committee about running an official candidate for the board of directors. However, during his interview, he refused to promise to not participate in any civil disobedience as part of his environmental activism, and the committee ultimately chose not to nominate him.

68. See Broydo, "Mutiny at the Sierra Club."

69. St. Clair and Cockburn, "Whither the Sierra Club?"

70. Ibid.

71. Chad Hanson interview.

72. Andy Mahler interview.

73. Dowie, *American Foundations*, 100. Other large foundations such as the Ford Foundation used their grants to promote collaborative logging projects in which community members living near national forests would forgo some of their enforcement of forest protection laws in order to allow what they hoped would be more ecologically appropriate logging projects on national forests. Despite the good intentions of their founders, these collaborative projects risked becoming a vehicle for timber companies to increase their access to national forests. The best-known example of this danger was a piece of federal legislation known as the Quincy Library Group Act, which promoted more logging on national forests in the Sierra Nevada mountain range of California. Although opposed by most environmental organizations, the act became law with the backing of Senator Dianne Feinstein (D-CA), who had previously been considered an ally of the environmental movement. The Quincy Library Group Act was an important precursor to the pattern of exempting national forest logging from oversight under the pretext of protecting forest health, which became more widespread during the George W. Bush administration.

74. Up until this time, the Western Ancient Forest Campaign (later renamed the American Lands Alliance) had served as the main voice for grassroots forest groups in Washington, DC, but its board was dominated by groups that were involved in the implementation of Clinton's Option 9 forest plan and wary of zero cut. In this context, zero-cut activists felt that it was more productive to form their own coalition. American Lands eventually endorsed the NFPRA and also provided an office for Rene Voss of the John Muir Project, but zero cut was not one its primary goals.

75. They hired the consulting firm of Michael Shellenberger, the future coauthor of "The Death of Environmentalism." However, the zero-cut activists came away deeply disappointed with Shellenberger's costly strategic planning report.

76. Hughes, "GOP Sees Foundations as the Menace in Forest Lands Restrictions."

77. Charlotte Talberth interview.

78. Quoted in Martin, "Clinton's Plan to Protect U.S. Forests," A1.

79. Douglas Jehl, "Road Ban Set for One-Third of U.S. Forests."

80. Although the Heritage Forests Campaign helped generate more than 1 million letters and e-mails from the public, that feedback did not dramatically change the roadless policy. Indeed, policy analyst Sheldon Kamieneki concluded that the final version of the Roadless Rule was slightly less strict than the proposed rule. Although the final version added the Tongass National Forest (albeit after a five-year delay that grandfathered in timber sales there), it also contained language that potentially could be used as a loophole to allow further logging and even road building in the roadless areas. See Kamieneki, *Corporate America and Environmental Policy*.

81. See, for example, Power, *Lost Landscapes and Failed Economies*; and Freudenburg, Wilson, and O'Leary, "Forty Years of Spotted Owls?"
82. See Wuerthner, ed., *Wildfire*.
83. Ibid.
84. Chad Hanson, "Big Timber's Big Lies."
85. Hanson, *Ending Logging on National Forests: The Facts in the Year 2000*.
86. Vaughn and Cortner, *George W. Bush's Healthy Forests*.
87. Quoted in St. Clair, *Been Brown So Long It Looked Like Green to Me*, 262.
88. Quoted in Vaughn and Cortner, *George W. Bush's Healthy Forests*, 152.
89. Quoted in St. Clair, *Been Brown So Long It Looked Like Green to Me*, 259–260.
90. For a detailed legislative history of the Healthy Forests Restoration Act, see Vaughn and Cortner, *George W. Bush's Healthy Forests*.
91. U.S. Forest Service, "Cut and Sold Reports."
92. Quoted in Dowie, *Losing Ground*, 218–19.
93. Even the northern spotted owl case, which involved significant participation by national organizations, can be traced to the impetus of grassroots activists, such as John Talberth's mass appeals of old-growth timber sales in Oregon and Max Strahan's Endangered Species Act listing petition for the owl, which were done despite opposition from the nationals.
94. Vaughn and Cortner, *George W. Bush's Healthy Forests*, 64. The Sierra Club's total included appeals filed by Sierra Club chapter-level activists. Vaugh and Cortner also note that the Forest Service did not systematically publish data on the appeals it received prior to 1997. The agency only began doing so as part of a court settlement in response to a lawsuit by an environmental group.

Chapter 5

1. Center for Biological Diversity, *Celebrating Fifteen Years of Protecting the Wild*, 1.
2. McGivney, "Moses or Menace?," 50.
3. Center for Biological Diversity, *Celebrating Fifteen Years of Protecting the Wild*, 1.
4. Kiefer, "Owl See You in Court," 22.
5. Yaffee, *Prohibitive Policy*, 48.
6. Quoted in Stanford Environmental Law Society, *The Endangered Species Act*, 21.
7. Quoted in Petersen, *Acting for Endangered Species*, ix.
8. Both endangered and threatened species are often collectively referred to as "endangered species." For brevity's sake, I adopt that practice in this chapter, making specific references to threatened species only when that distinction is relevant. That said, it should be noted that, in practice, threatened species do not receive the same level of protection under the ESA as endangered species.
9. The National Marine Fisheries Service is tasked with implementing the ESA for marine species such as salmon, whales, and sea turtles.
10. There are four exceptions to the prohibition on take: (1) if an incidental take permit is granted by the FWS to private parties under Section 10 of the ESA; (2) if an incidental take statement is granted to other government agencies under Section 7; (3) in some cases, take can be allowed for threatened species, pursuant to regulations under what is called a 4(d) rule; and (4) in some cases, endangered plant species on private lands can be killed without a permit.
11. Quoted in Stanford Environmental Law Society, *The Endangered Species Act*, 106.
12. Ibid., 202.
13. In the mid-1970s, the director of the Fish and Wildlife Service told Congress that the agency intended to list about 3,000 animals and 4,000 plants by 1982. In reality though, the agency never came close to achieving that goal. See

Richard Tobin, *The Expendable Future*.

14. Yaffee, *Prohibitive Policy*.

15. Plater, "Endangered Species Act Lessons over 30 Years and the Legacy of the Snail Darter."

16. Quoted in Petersen, *Acting for Endangered Species*, 55.

17. Yaffee, *Prohibitive Policy*, 136–37.

18. The 1978 amendment also allowed the agency to consider economic impact when determining the critical habitat for a species, though not during the decision to list the species. At the same time, the amendment contained some less-noticed provisions to strengthen the ESA, including the requirement that critical habitat for a species be designated at the same time the species is listed. There were also new requirements to develop and implement recovery plans for the listed species and increased funding for endangered plant protection.

19. Prior to Baker's rider, Al Gore Jr., then a congressman for Tennessee, had unsuccessfully introduced legislation to exempt the Tellico dam from the ESA.

20. See Peterson, *Acting for Endangered Species*.

21. Greenwald, Suckling, and Taylor, "The Listing Record," 56.

22. Ibid., 55.

23. In these amendments, Congress reaffirmed that listing decisions should be without regard to economic issues. The amendments also contained a provision that allowed incidental take permits to be issued in exchange for mitigation measures to increase species protection elsewhere (at least in theory) through a system of habitat conservation plans. This provision was rarely used by the Reagan and Bush administrations in the 1980s, but it became a defining feature of Clinton's ESA policies in 1990s.

24. U.S. Department of the Interior Office of Inspector General, *Audit Report: The Endangered Species Program, U.S. Fish and Wildlife Service*, i, 7.

25. The ESA was amended again in 1988. The amendments included minor provisions to strengthen the ESA but did not have a significant influence on the implementation of the act. As of this writing, the 1988 legislation was the last time that the ESA was amended by Congress.

26. As mentioned in the preface, Jasper Carlton has faced serious health concerns and had to step away from environmental activism in the early 2000s. As a result, he could not be interviewed for this book. Instead, much of the material in this section comes from interviews with activists who worked closely with him.

27. Ned Mudd interview.

28. According to some accounts, Carlton may have previously played a role in the effort to protect the snail darter, the foundational example of grassroots ESA litigation.

29. Halverson, "A Full-Court Press to Save Ecosystems."

30. The woodland caribou was later one of the species highlighted in a U.S. Postal Service stamp series commemorating endangered species protection.

31. Young, "Abracadabra!"

32. For a profile of Ned Mudd, see Little, "Alabama's Green-Leanin' Good Ol' Boys."

33. Anne Becher, *American Environmental Leaders*.

34. Ned Mudd interview.

35. Jasper Carlton had proposed ten lawsuits for which he had not been able to find a lawyer. When Eric Glitzenstein offered to take on the massive multispecies suit, Carlton initially stormed out of the meeting declaring that Glitzenstein was not committed enough to the issue if he would only do one case. Glitzstein fondly

recounted this incident as indicative of Carlton's fierce determination. His intense focus on endangered species protection alienated some people, but that same commitment also attracted allies such as Glitzenstein to work on Carlton's cases.

36. Eric Glitzenstein interview.
37. By then Eric Glitzenstein had left the Public Citizen Litigation Group. He and Kathy Meyer would later start Meyer & Glitzenstein, one of the most prominent public interest law firms for endangered species litigation.
38. Kieran Suckling interview.
39. For a more detailed discussion of the Center's three-step strategy, see Aleshire, "A Bare-Knuckled Trio."
40. Quoted in Aleshire, "A Bare-Knuckled Trio," 10–11.
41. Murphy, "Timber Firms Endangered by Owl Ruling," B1.
42. For the sake of continuity, I use the shorter form of the Center's name without "Southwest" when discussing the group during its early years as well as after the official name change took place in 1999.
43. Carson Forest Watch, Forest Conservation Council, Forest Guardians, Maricopa Audubon, and Dine CARE, a Navajo environmental group.
44. Quoted in Kiefer, "Owl See You in Court," 24.
45. Zakin, "The Wild Bunch."
46. U.S. Forest Service, "Cut and Sold Reports."
47. The Mexican spotted owl injunction was strongly criticized by Arizona's governor and other prominent figures, but public support for the injunction remained high. At the time, Earthlaw attorney Mark Hughes noted "the results of a poll taken in Arizona's most populous county. The poll shows that most of those surveyed agreed with the judge's order to halt logging. This support crosses party and age lines. Fifty-one percent of Republicans, 70 percent of Democrats, and 75 percent of independents agreed with the judge's order. A majority of those over 55 agreed, and among those 18 to 25 years old, support for the injunction is 88 percent. The latter figures led the pollsters to the encouraging conclusion that 'support for these types of environmental protection measures will probably grow.'" See Hughes, "Southwest Logging Ban Sustained."
48. Quoted in Aleshire, "A Bare-Knuckled Trio," 8.
49. U.S. Forest Service, "Cut and Sold Reports."
50. Quoted in Walley, "Playing Outside the Rules."
51. Southwest Center for Biological Diversity, *Annual Report 1997*, 4.
52. Center for Biological Diversity, *Celebrating Fifteen Years of Protecting the Wild*, 1.
53. Davis, "Environmental Groups File Suit in Owl Dispute," 1B.
54. Davis, "County Alters Open-Space Guides," 2B.
55. Bagwell, "Pygmy Owl Champions Crusade for Humanity."
56. Tom Kenworthy, "In Desert Southwest, a Vigorous Species Act Endangers a Way of Life," A3.
57. Bagwell, "Group Loses Suit to Save Lake Mead Birds."
58. Southwest Center for Biological Diversity, *Annual Report 1997*, 9.
59. Lemann, "No People Allowed."
60. David Hogan interview.
61. McGivney, "Moses or Menace?," 53.
62. Kieran Suckling interview.
63. Aleshire, "Can Southwest Activism and Money Coexist?"
64. Southwest Center for Biological Diversity, *Annual Report 1996*.
65. Suckling, "Some Thoughts on Big Money and Bad Activism."
66. Ibid., 8.

67. The ten organizations represented on the ESC steering committee were Defenders of Wildlife, Environmental Defense Fund, Greenpeace, Humane Society of the United States, National Audubon Society, National Wildlife Federation, Sierra Club, Sierra Club Legal Defense Fund, The Wilderness Society, World Wildlife Fund, with the Center for Marine Conservation and Natural Resources Defense Council having advisory status.
68. Strickler, "The Do Nothing Strategy."
69. Ibid.
70. As noted earlier, during that time, the Center and its coplaintiffs were able to get the Hutchinson rider negated with regard to the Mexican spotted owl case.
71. See, for example, Dowie, *American Foundations*.
72. Ibid., 101.
73. Medberry, "Who Knows Best: Grassroots or Foundations?"
74. The National Environmental Trust ultimately merged with Pew in 2007 under the name of the Pew Environmental Group, as Pew's environmental program changed from a traditional grantmaking role into becoming its own advocacy organization.
75. Dowie, *Losing Ground*.
76. Bergoffen, "Coalition to Environmental Defense Fund: Get Out!"
77. Michael Bean interview.
78. In regard to the other three groups involved in the secret negotiations: the Center for Marine Conservation (later renamed the Ocean Conservancy) resigned from the ESC soon after EDF departed; the World Wildlife Fund received a letter of censure from the ESC; and the Nature Conservancy was not a member of the ESC.
79. Subsequent to this interview, Bill Snape began working for the Center for Biological Diversity in Washington, DC.
80. McClure and Stiffler, "Flaws in Habitat Conservation Plans Threaten Scores of Species."
81. Ibid.
82. American Lands Alliance, "Broken Promises of Recovery."
83. McClure and Stiffler, "Lone Voice Challenges 'No Surprises.'"
84. Babbitt also stated that, if he had been given the option, he would not have approved the sweeping settlement by the Bush administration of Jasper Carlton's case regarding candidate species. See Noah, "Caught in a Trap."
85. Quoted in Greenwald, Suckling, and Taylor, "The Listing Record," 60.
86. Wilson, "Enforcement of Species Act Pits Babbitt vs. Activists," B1.
87. Brendan Cummings interview.
88. Suckling and Taylor, "Critical Habitat and Recovery," 76.
89. Center for Biological Diversity, *Celebrating Fifteen Years of Protecting the Wild*, 1.
90. The hiring of Jamie Rappaport Clark marked a key shift in Defenders of Wildlife. In the 1990s and early 2000s, Defenders of Wildlife was sometimes willing to be a coplaintiff in litigation with grassroots biodiversity groups. During that time, Clark, as director of the Fish and Wildlife Service, was sued repeatedly by these groups for her agency's failure to implement the ESA. Despite the willingness of some staff to work more closely with groups like the Center, Defenders of Wildlife as a whole still operated on an insider strategy like the other nationals. And from the perspective of an insider strategy, hiring a well-connected Clinton administration official like Clark as a vice president of Defenders made sense to the organization's leadership. Yet, after she was hired, Defenders' endangered species policies became more closely aligned with the Environmental

Defense Fund rather than grassroots groups. In this regard, the experiences of Defenders illustrates that while there can be variation in the behavior of different national organizations, ultimately the actions of all of those organizations are circumscribed by their reliance on an insider strategy.

91. Quoted in Suckling and Taylor, "The Listing Record," 78 (italics added).

92. Greenwald, Suckling, and Taylor, "The Listing Record."

93. Wilson, "Enforcement of Species Act Pits Babbitt vs. Activists."

94. Kassie Siegel would later write the ESA listing petition for the polar bear. See chapter 6.

95. Up until the early 2000s, the Center did not promote specialized roles within the organization. Even the early staff attorneys were not limited to working on litigation. Instead they acted like other Center campaigners in terms of their responsibilities for working on the science-based listing petitions, talking to the media, and other aspects of the campaign.

96. Center for Biological Diversity, *Annual Report 2000*.

97. Jay Tutchton interview.

98. A later analysis of references to environmental organizations in news articles found that the Center for Biological Diversity received the fifth highest level of news coverage of any environmental organization. Moreover, when the total number of news appearances by an organization was divided by that group's budget, the analysis showed that the Center received a greater amount of coverage per dollar than any other organization. See Suckling, *Environmental Media Generation Statistics*.

99. Center for Biological Diversity, *Annual Report 2004*.

100. Center for Biological Diversity, *Celebrating Fifteen Years of Protecting the Wild*.

101. Rogers, *Acorn Days*.

102. Fox, *American Conservation Movement*.

103. Dowie, *Losing Ground*, 37–38.

104. Dowie, *Losing Ground*.

105. While having a board of founding staff may be beneficial in terms of organizational mission, it can also present difficulties in terms of organizational decision making. Because the founding staff are also the primary board members, they have ultimate authority over the organization. Therefore there is a sharp inequality between the level of decision-making power of the founding staff and the rest of the staff. While this problem may grow as the organization continues to expand, for much of the Center's history, the issue has been mitigated by giving the staff a high degree of autonomy, thus making these differences in power less important in practice.

106. Gottlieb, *Forcing the Spring*.

107. McGivney, "Moses or Menace?," 53.

108. With the addition of these new staff members, the number of women working at Center grew to almost 50 percent by 2004.

109. Professionalization alone was not entirely responsible for this shift. Instead, Kieran Suckling attributed the shift to the waning influence of Earth First!. He noted that initially Earth First! had been the sole refuge for activists who were dissatisfied with moderate environmentalism. But with the development of radical grassroots groups such as the Center, young activists who in the past might have first participated in Earth First! instead went directly into internships with the Center. Earth First! thus became less important as a labor pool for the Center. This also meant that Earth First! began to play less of a role in nurturing the radicalism of the Center's staff. While participation in Earth First! was not the

only path to radicalism for environmental activists, it had been a reliable one. The experience of direct action and the free flow of unconventional ideas around the campfires at Earth First! gatherings had a transformative effect. It is not clear whether the Center on its own can provide similar radicalizing experiences for young activists through working in an office on listing petitions and fundraising appeals.

110. Despite the successes that accompanied the Center's unconventional approach to its organizational structure, cofounder Todd Schulke expressed concern that "there is that potential—more out of fatigue than anything—of being pushed into a more conventional structure that might not allow us to be the way we have been." As an example, Schulke cited the influence of Training Resources for the Environmental Community (TREC). In the early 2000s, the Wilburforce Foundation, an environmental grantmaker that supports many grassroots groups, funded the creation of TREC as a service to its grantees. TREC offered a team of career nonprofit administrators to provide training for those environmental groups in professional organizational management. As a Wilburforce grantee, the Center participated in TREC's leadership training program and undertook some organizational restructuring during this time. But Schulke raised concerns about the homogenizing effect of the TREC professionalization process for the Center and other grassroots groups:

> I don't want to be too cynical about what I call the TREC model because I know it helps a lot of people get organized and get less dependent on foundations. But we have operated differently than any other group I know of, and I don't know that there's a map for us. Right now almost every group I know of is going through these TREC development programs, so we're all going to be the same. And that's not a good thing. I have great respect for the TREC folks. I just don't think it is good to have this leveling effect that everybody's structure is all going to be the same. The fact that we have done things dramatically different from almost every other group in the country has to be important.

Chapter 6

1. Quoted in Walley, "Playing outside the Rules."
2. In this regard, grassroots biodiversity litigation is similar to the industrial workers' sit-down strikes of the 1930s, which advanced unionization, and the civil rights movement sit-ins in the early 1960s, which propelled the end of segregation in the South. As social movement scholars Frances Fox Piven and Richard Cloward highlighted in *Poor People's Movements*, these earlier tactics worked particularly well because they did not require an extensive organization or resources in order to have a big impact. Thus, the activists were able to exert influence while avoiding the constraints that Piven and Cloward identified with large social movement organizations working through conventional political processes. While similar in terms of its decentralized nature, grassroots biodiversity litigation differs from the earlier tactics in that sit-ins and sit-down strikes were illegal and thus vulnerable to police repression. Thus, litigation is a potentially a more sustainable tactic.
3. To be clear, Earth First! was not the only path to becoming a radical grassroots activist, though it was a fairly reliable route. Another notable route was embodied by Sierra Club volunteers such as Tim Hermach, Jim Bensman, and Chad Hanson, who became radicalized through their frustration with the constraints on forest protection advocacy imposed by the Club's staff and leadership.

4. However, in the late 1990s and 2000s, Earth First!'s influence began to fade. In part, this shift was a result of the success of the new grassroots groups. Previously, Earth First! had been the sole refuge for activists who were dissatisfied with moderate environmentalism. But as new grassroots groups accumulated accomplishments during the 1990s, young activists who in the past might have otherwise participated in Earth First! instead went directly into working with these groups. However, most of the new groups did not develop their own radicalizing, movement-building institutions, and it is not clear whether simply working on timber sale appeals and ESA listing petitions can have the same radicalizing effect on a new generation of activists as participation in Earth First! did for the previous generation. Furthermore, by the early 2000s, the founders of the grassroots groups were less likely to attend Earth First! gatherings or read the *Earth First! Journal* on a regular basis. Having a background in Earth First! does not guarantee life-long use of an outsider strategy. There are ongoing inducements for outsider groups to adopt a more conventional approach to social change, and engaging in contentious conflicts can be exhausting. It is not surprising then that some grassroots activists became more conflict-averse over time and later adopted an insider approach. Ultimately, it is only through the ongoing encouragement of a radical movement culture that an outsider strategy is likely to be sustained. The decline of this radical movement culture emerged as an important challenge for grassroots biodiversity groups in the 2000s.
5. A similar phenomenon may develop when numerous small organizations are amalgamated within a larger coalition, especially when that coalition has shared funding, such as in the case of the Headwaters Forest Coordinating Committee. In this regard, the preference of some grantmakers to fund formal campaign coalitions may actually be detrimental to those campaigns.
6. Quoted in Dowie, *American Foundations*, 100.
7. In one notable exception, the Center for Biological Diversity rapidly expanded its membership fundraising in the early 2000s, but that development was possible only after a decade of foundation funding allowed the Center to grow large enough and generate enough media attention to make direct mail fundraising economically viable.
8. The importance of movement activists occupying staff positions in foundations for the development of more radical grassroots groups is paralleled in sociologists Daniel Faber and Deborah McCarthy's study of the environmental justice movement. See Faber and McCarthy, "Breaking the Funding Barriers."
9. Quoted in Bevington, "Earth First! in Northern California," 270.
10. President Clinton was also more likely to appoint federal judges who were more likely to enforce environmental laws than George W. Bush's choices. That said, it should be noted that Judge Dwyer, who was responsible for the crucial northern spotted owl injunctions, was a Reagan appointee.
11. Kamieniecki, Coleman, and Vos, "The Effectiveness of Radical Environmentalists," 324–25.
12. Quoted in Dowie, *Losing Ground*, 218–19.
13. During the time period covered by this study, some national organizations did play a prominent role in getting some areas added to wilderness status through legislation. Wilderness legislation was particularly compatible with the political deal making of the insider strategy, with the protection of some wildlands traded away in exchange for securing the inclusion of other areas. However, these wilderness bills had little direct impact on the forest and endangered species protection campaigns described in this book.

14. Dowie, *Losing Ground*, 248.
15. Ibid., 250.
16. The Fund for Wild Nature (see chapter 2) has became a key resource for linking small-scale donors to grassroots groups. While most other environmental foundations were created through an endowment from a wealthy donor, the Fund for Wild Nature instead relies on annual contributions from the public, which it then redistributes as grants to grassroots biodiversity groups throughout the United States.
17. Thomas et al., "Extinction Risk from Climate Change."
18. Gelbspan, *Boiling Point*, 128.
19. Ibid., 133.
20. Dorsey, "Carbon Trading Won't Work."
21. Wilkerson, "Climate Activists Invade DC Offices of Environmental Defense Fund."
22. Wayburn et al., *Forest Carbon in the United States*, 3.
23. Charlotte Talberth of Levinson Foundation later became one of the first funders to sponsor the Center's emerging climate program. As with her earlier support of the zero-cut campaign, Talberth undertook a grassroots funding strategy for addressing the climate crisis.
24. *Time*, April 3, 2006.

Afterword

1. Suckling, "Ken Salazar a Disappointing Choice for Secretary of the Interior."
2. Pace, "Green Myopia."

Appendix

1. Gottlieb, *Forcing the Spring*, 124.
2. Bean and Rowland, *The Evolution of National Wildlife Law*, 123.
3. Weber, *From Abundance to Scarcity*, 125.
4. Ibid.
5. Reffalt, "Endangered Species Lists," 78.
6. Quoted in Stanford Environmental Law Society, *The Endangered Species Act*, 20.
7. See Weber, *From Abundance to Scarcity*.
8. Ibid., 52.
9. Hirt, *A Conspiracy of Optimism*, 246.
10. Ibid., 261.
11. Ibid., 262.
12. LeMaster, *Decade of Change*, 64.

GLOSSARY OF ACRONYMS

ASCMEE. Association of Sierra Club Members for Environmental Ethics

BACH. Bay Area Coalition for Headwaters

BLM. Bureau of Land Management

CDF. California Department of Forestry and Fire Protection

EDF. Environmental Defense Fund

EPIC. Environmental Protection Information Center

ESA. Endangered Species Act

ESC. Endangered Species Coalition

FBI. Federal Bureau of Investigation

FDIC. Federal Deposit Insurance Corporation

FWS. U.S. Fish and Wildlife Service

HCP. Habitat conservation plan

HFC. Heritage Forests Campaign

HFCC. Headwaters Forest Coordinating Committee

JMS. John Muir Sierrans

NFC. Native Forest Council

NFPRA. National Forest Protection and Restoration Act

ONRC. Oregon Natural Resources Council

OTS. Office of Thrift Supervision

OWC. Oregon Wilderness Coalition

PL. Pacific Lumber Company

POW. Protect Our Woods

RARE II. Roadless Area Review and Evaluation II

TREC. Training Resources for the Environmental Community

WTO. World Trade Organization

BIBLIOGRAPHY

Abate, Thomas. "Lumber Dispute Coming to a Head: State Water Panel to Rule on Harvesting." *San Francisco Chronicle*, June 12, 2005.

———. "Hurwitz Wins FDIC Lawsuit: Bank Regulatory Agency Is Ordered to Repay $72 Million." *San Francisco Chronicle*, August 25, 2005.

Aleshire, Peter. "Can Southwest Activism and Money Coexist?" *High Country News*, April 15 1996. http://www.hcn.org/servlets/hcn.Article?article_id=1774 (accessed March 1, 2005).

———. "A Bare-Knuckled Trio Goes after the Forest Service." *High Country News*, March 30, 1998.

American Lands Alliance. "Broken Promises of Recovery: The Clinton Administration's 10-Prong Attack on Endangered Species." http://www.americanlands.org/documents /1091742884_brokproms.pdf (accessed March 1, 2005).

Apple, Agent [pseud.], ed. *Pie Any Means Necessary: The Biotic Baking Brigade Cookbook.* Oakland: AK Press, 2004.

Arquilla, John, and David Ronfeldt, eds. *Networks and Netwars.* Santa Monica: RAND, 2001.

Bader, Mike. "Northern Rockies Ecosystem vs. Washington DC Political System." *Wild Earth,* Fall, 1991.

Bagwell, Keith. "Group Loses Suit to Save Lake Mead Birds." *Arizona Daily Star,* June 21, 1997.

———. "Pygmy Owl Champions Crusade for Humanity: Environmental Group Winning War in Court." *Arizona Daily Star,* December 7, 1997. http://www.endangeredearth.org/alerts /result.asp?index=274 (accessed April 1, 2005).

Bailey, Kathy. "Timber Accord: What's the Big Deal?" *Santa Rosa Press Democrat,* July 20, 1992.

———. "Why I Should Feel Good (Even If I Don't)." unpublished memo, 1999.

Ball, Gary. "State Forestry Update." Mendocino Environmental Center. http://www.mecgrassroots.org/NEWSL/ISS10/10.03state.html (accessed March 1, 2005).

Bari, Judi. *Timber Wars.* Monroe: Common Courage Press, 1994.

———. "FBI Bomb School and Other Atrocities." *Anderson Valley Advertiser,* October 19, 1994.

Barnum, Alex. "Protests Widen over Headwaters: Activists Blockade Roads, Call Negotiations 'a Sham.'" *San Francisco Chronicle,* September 17, 1996.

———. "Toppled Tree Kills Logging Protester in Humboldt: First Fatality in Decade of Activism." *San Francisco Chronicle,* September 18, 1998.

Barnum, Alex, and Elliot Diringer. "Juggling Trees, Wildlife and Property Rights: The Headwaters Deal, with Primeval Forests at Stake, is a High-Profile Test of the Country's Conservation Efforts." *San Francisco Chronicle,* October 13, 1996.

Beach, Patrick. *A Good Forest for Dying: The Tragic Death of a Young Man on the Front Lines of the Environmental Wars.* New York: Doubleday, 2003.

Bean, Michael, and Melanie Rowland. *The Evolution of National Wildlife Law.* 3rd. ed. Westport: Praeger, 1997.

Bergoffen, Marty. "Coalition to Environmental Defense Fund: Get Out!" *Earth First! Journal* 16(5), 1996.

Becher, Anne. *American Environmental Leaders: From Colonial Times to the Present.* Vol. 1, *A-K.* Santa Barbara: ABC-CLIO, 2000.

Bevington, Douglas. "Lessons from the HFCC in Campaign Coalition Organization." unpublished memo, 1997.

———. "Earth First! in Northern California: An Interview with Judi Bari." In *The Struggle for Ecological Democracy: Environmental Justice Movements in the United States*, edited by Daniel Faber, 248–71. New York: Guilford Press, 1998.

Bevington, Douglas, and Chris Dixon. "Movement-Relevant Theory: Rethinking Social Movement Scholarship and Activism." *Social Movement Studies*, 4(3) (2005): 185–208.

Broswimmer, Franz. *Ecocide: A Short History of Mass Extinction of Species*. Sterling: Pluto Press, 2002.

Brower, David. *For Earth's Sake: The Life and Times of David Brower*. Salt Lake City: Peregrine Smith, 1990.

———. *Let the Mountains Talk, Let the Rivers Run: A Call to Those Who Would Save the Wild*. San Francisco: Harper San Francisco, 1995.

Broydo, Leora. "Mutiny at the Sierra Club." *Mother Jones*, November 3. 1998. http://www.motherjones.com/news/feature/1998/11/sierra.html (accessed March 1, 2006).

Brulle, Robert. *Agency, Democracy, and Nature: The U.S. Environmental Movement from a Critical Theory Perspective*. Cambridge: MIT Press, 2000.

Bundy, Kevin. "The People's History of Activism from Headwaters." *Headwaters Forest Update*, Summer, 1999.

California Public Employees for Environmental Responsibility. *California's Failed Forest Policy: State Biologists Speak Out*. Sacramento: California Public Employees for Environmental Responsibility, 2000.

Center for Biological Diversity. *Celebrating Fifteen Years of Protecting the Wild*. Tucson: Center for Biological Diversity, 2004.

———. *Center for Biological Diversity Annual Report 2004*. Tucson: Center for Biological Diversity, 2004.

Champion, Dale. "Redwood Cutting Plan Provokes a Protest." *San Francisco Chronicle*, October 23, 1986.

Cherney, Darryl. "If It Ain't Broke, Don't Fix It." *Earth First! Journal* 16(3), 1996.

Chronicle Staff Report. "Headwaters Protest Turns Ugly: Environmentalists Trash Demos' Office—Feinstein Furious." *San Francisco Chronicle*, October 1, 1996.

Clifford, Frank. "Deal to Save the Headwaters Could Fail." *Los Angeles Times*, December 18, 1998.

Cohen, Michael. *The History of the Sierra Club, 1892–1970*. San Francisco: Sierra Club Books, 1988.

Daily, Gretchen, ed. *Nature's Services: Societal Dependence on Natural Ecosystems*. Washington, DC: Island Press, 1997.

Davis, Tony. "Environmental Groups File Suit in Owl Dispute." *Arizona Daily Star*, December 20, 1997.

———. "County Alters Open-Space Guides." *Arizona Daily Star*, June 22, 2005.

Devall, William. "The Governing of a Voluntary Organization: Oligarchy and Democracy in the Sierra Club." PhD diss., University of Oregon, 1970.

Devall, Bill, and George Session. *Deep Ecology: Living as if Nature Mattered*. Salt Lake City: Peregrine Smith, 1985.

Dorsey, Michael. "Carbon Trading Won't Work." *Los Angeles Times*, April 1, 2007. http://www.latimes.com/news/opinion/sunday/commentary/la-op-dorsey1apr01,0,7611817 .story?coll=la-sunday-commentary (accessed on May 1, 2007).

Dowie, Mark. *Losing Ground: American Environmentalism at the Close of the Twentieth Century*. Cambridge: MIT Press, 1995.

———. *American Foundations: An Investigative History*. Cambridge: MIT Press, 2001.

Driscoll, John. "State Water Board: PL's Peril Its Own Making." *Eureka Times-Standard*, May 7, 2005. http://www.epochdesign.com/hwc/Press/TS%20050705.htm (accessed August 1, 2005).

Durbin, Kathie. *Tree Huggers: Victory Defeat and Renewal in the Northwest Ancient Forest Campaign*. Seattle: Mountaineers, 1996.

Environmental Protection Information Center. "Summary of EPIC Forestry Lawsuits 1982 to 1996." http://www.wildcalifornia.org/litigation/litsum82_96.html (accessed March 1, 2005).

———. "EPIC's Historic Case Goes to Trial: Pacific Lumber's Permits to Kill Indicted." *Wild California*, Spring, 2003. http://www.wildcalifornia.org/publications/article-45 (accessed on March 1, 2005).

Examiner Editorial Board. "Salvation Is at Hand." *San Francisco Examiner*, September 3, 1998. http://www.sfgate.com/cgi-bin/article.cgi?f=/e/a/1998/09/03/EDITORIAL7561.dtl&hw =Headwaters&sn= 004&sc=430 (accessed March 1, 2005).

Faber, Daniel, and Deborah McCarthy, eds. *Foundations for Social Change: Critical Perspectives on Philanthropy and Popular Movements*. Lanham: Rowman & Littlefield, 2005.

———. "Breaking the Funding Barriers: Philanthropic Activism in Support of the Environmental Justice Movement." In *Foundations for Social Change: Critical Perspectives on Philanthropy and Popular Movements*, edited by Daniel Faber and Deborah McCarthy, 175–209. Lanham: Rowman & Littlefield, 2005.

Foreman, Dave. *Confessions of an Eco-Warrior*. New York: Harmony, 1991.

Fox, Stephen. *The American Conservation Movement: John Muir and His Legacy*. Madison: University of Wisconsin Press, 1981.

Freeman, Jo. "The Tyranny of Structurelessness." *Berkeley Journal of Sociology* 17 (1972): 151–64.

Freudenburg, William, Lisa Wilson, and Daniel O'Leary. "Forty Years of Spotted Owls? A Longitudinal Analysis of Logging-Industry Job Losses." *Sociological Perspectives* 41(1) (1998): 1–26.

Gelbspan, Ross. *Boiling Point: How Politicians, Big Oil and Coal, Journalists, and Activists Have Fueled the Climate Crisis—And What We Can Do to Avert Disaster*. New York: Basic Books, 2004.

Ghent, Randy. "Thanksgiving Massacres on West Coast: California's Owl Creek." *Earth First! Journal* 13(2), 1992.

———. "Mass Action for Headwaters: 264 Arrested with More to Follow . . . " *Earth First! Journal* 15(8), 1995.

———. "Don't Take Orders from Anyone." *Earth First! Journal* 16(2), 1995.

Gledhill, Lynda. "Davis' Final Bills Aid Environment: New Law Lets Water Districts Block Logging Plans." *San Francisco Chronicle*, October 14, 2003.

Goodman, John. "Standing Up for Democracy." In *Voices from the WTO: An Anthology of Writings from the People Who Shut Down the World Trade Organization*, edited by Stephanie Guilloud, 17. Olympia: Evergreen State College, 2000.

Gottlieb, Robert. *Forcing the Spring: The Transformation of the American Environmental Movement*. Washington, DC: Island Press, 1993.

Greenwald, Noah, Kieran Suckling, and Martin Taylor. "The Listing Record." In *The Endangered Species Act at Thirty*. Vol. 1, *Renewing the Conservation Promise*, edited by Dale Goble, J. Michael Scott, and Frank Davis, 51–67. Washington, DC: Island Press, 2006.

Gunnison, Robert, and Greg Lucas. "Critics Say Davis Kowtows to Donors: Access Being Sold, They Charge." *San Francisco Chronicle*, July 20, 1999.

Gurnon, Emily. "One Year Later." *North Coast Journal Weekly*. March 3, 2005. http://www.northcoastjournal.com/030305/cover0303.html (accessed August 1, 2005).

Halverson, Anders. "A Full-Court Press to Save Ecosystems." *High Country News*, May 15, 1995. http://www.hcn.org/servlets/hcn.PrintableArticle?article_id=1037 (accessed April 30, 2006).

Hanson, Chad. *Ending Logging on National Forests: The Facts*. Pasadena: John Muir Project of Earth Island Institute, 1997.

———. *Ending Logging on National Forests: The Facts in the Year 2000*. Pasadena: John Muir Project of Earth Island Institute, 2000.

———. "Big Timber's Big Lies: They'll Say Whatever It Takes to Keep the Subsidies Rolling In." *Sierra*, September/October, 2000. http://www.sierraclub.org/sierra/200009/timber.asp (accessed March 1, 2006).

Harris, David. *The Last Stand: The War between Wall Street and Main Street over California's Ancient Redwoods*. New York: Times Books, 1995.

Higgs, Steven. *Eternal Vigilance: Nine Tales of Environmental Heroism in Indiana*. Bloomington: Indiana University Press, 1995.

Hill, Julia Butterfly. *The Legacy of Luna: The Story of a Tree, a Woman, and the Struggle to Save the Redwoods*. San Francisco: Harper San Francisco, 2000.

Hirt, Paul. *A Conspiracy of Optimism: Management of the National Forests since World War Two*. Lincoln: University of Nebraska Press, 1994.

Hughes, John. "GOP Sees Foundations as the Menace in Forest Lands Restrictions." Associated Press, February 24, 2000. http://web.lexis- nexis.com/universe/document?_m =c01856b1bba62e4bf42a8828552de275&_docnum=14&wchp=dGLzVlz-zSkVA&_md5=d4e00ace6a7f9801578f4ff3294b825d (accessed March 1, 2006).

Hughes, Mark. "Southwest Logging Ban Sustained—You Can't Tell a Book by Its Cover." *Earthlaw Newsletter*, August 6, 1996. http://www.law.du.edu/hughes/Elaw/Feature/Newslett/memo10.htm (accessed on June 1, 2007).

Earth First! Journal. 1998a. "If Words Could Kill. . ." [transcript of the videotape of the confrontation between A.E. Ammons and David "Gypsy" Chain, September 17, 1997]. *Earth First! Journal*, Vol. 19(1), p.19.

INCITE! Women of Color against Violence, eds. *The Revolution Will Not Be Funded: Beyond the Non-Profit Industrial Complex*. Cambridge: South End Press, 2007.

Ingalsbee, Timothy. "Earth First!: Consciousness in Action in the Unfolding of a New-Social-Movement." PhD diss., University of Oregon, 1995.

Izakson, Orna. "Animal Rights: Legal Activists Fight for Endangered Species." *E: The Environmental Magazine*, May/June, 2002.

Jakubal, Mike. "Green Steel: Should Earth First! Work with the Steel Union?—Good Idea." *Earth First! Journal*, 19(5), 1999.

Jehl, Douglas. "Road Ban Set for One-Third of U.S. Forests." *New York Times*, January 5, 2001. http://www.nytimes.com/2001/01/05/politics/05ROAD.html?pagewanted=all (accessed March 1, 2006).

Kamieniecki, Sheldon. *Corporate America and Environmental Policy: How Often Does Business Get Its Way?* Stanford: Stanford University Press, 2006.

Kamieniecki, Sheldon, S. Dulaine Coleman, and Robert Vos. "The Effectiveness of Radical Environmentalists." In *Ecological Resistance Movements: The Global Emergence of Radical and Popular Environmentalism*, edited by Bron Taylor, 315–33. Albany: State University of New York Press, 1995.

Kay, Jane. "Fight for Old Growth Hasn't Ended." *San Francisco Examiner*, October 1, 1996. http://www.sfgate.com/cgi-bin/article.cgi?f=/e/a/1996/10/01/NEWS12083.dtl&hw =Kay&sn=004&sc=803 (accessed March 1, 2005).

Kegley, Don. "Victory! Victory! Victory!" *Green Worker* 1(1), 2000.

Kenworthy, Tom. "In Desert Southwest, a Vigorous Species Act Endangers a Way of Life." *Washington Post*, February 1, 1998.

Kiefer, Michael. "Owl See You in Court." *Phoenix New Times*, August 1, 1996.

———. "A Hostile Environment: Witnesses Say a Justice Department Attorney Physically Attacked a Lawyer for Environmentalists." *Phoenix New Times*, October 31, 1996.

http://promotions.phoenixnewtimes.com/1996-10-31/news/a-hostile-environment/ (accessed March 1, 2005).

Knickerbocker, Brad. "Debt-for-Nature Swap Proposed between Uncle Sam and Loggers." *Christian Science Monitor*, December 3, 1993.

Langelle, Orin. "Shawnee Showdown!" *Earth First! Journal* 11(8), 1991.

Lee, Jessica. "State Legislature Debates while Treesitters Hold Strong." *Earth First! Journal* 24(6), 2004.

Lee, Martha. *Earth First!: Environmental Apocalypse*. Syracuse: Syracuse University Press, 1995.

Lemann, Nicholas. "No People Allowed: A Radical Environmental Group Attempts to Return the Southwest to the Wild." *New Yorker*, November 22, 1999.

LeMaster, Dennis. *Decade of Change: The Remaking of Forest Service Statutory Authority during the 1970s*. Westport: Greenwood Press, 1984.

Lifsher, Marc."Lawyer Goes against the Grain in Logging Fight with Maxxam." *Wall Street Journal*, February 10, 1999.

Little, Jane. "Alabama's Green-Leanin' Good Ol' Boys," *Forest Magazine*, November/December, 1999. http://www.fseee.org/index.html?page=http%3A//www.fseee.org/forestmag/9901little.shtml (accessed June 1, 2005).

Lucas, Greg. "Timber Report Called Faulty: Ex-Forestry Official Says Pacific Lumber Understated Damage." *San Francisco Chronicle*, February 5, 2004.

Martin, Glen. "Clinton's Plan to Protect U.S. Forests: No Logging or Mining on 40 Million Acres, including 4 Million Acres in California." *San Francisco Chronicle*, October 14, 1999.

———. "D.A. Hits Timber Firm with Fraud Charges: Humboldt County Sues Pacific Lumber." *San Francisco Chronicle*, February 26, 2003.

Martin, Glen, and Jonathan Curiel. "Last-Minute Headwaters Deal OKd: Stands of Ancient Redwoods Preserved in Landmark Sale." *San Francisco Chronicle*, March 2, 1999.

Martin, Glen, Carolyn Lochhead, and Lynda Gledhill. "Headwaters Deal Lauded as Model Agreement: Pacific Lumber Co. Signs Off on Logging Pact Minutes Before Deadline." *San Francisco Chronicle*, March 3, 1999.

McClure, Robert, and Lisa Stiffler. "Flaws in Habitat Conservation Plans Threaten Scores of Species." *Seattle Post-Intelligencer*, May 3, 2005. http://seattlepi.nwsource.com/specials/licensetokill/222273_enviro03.html (accessed August 1, 2005).

———. "Lone Voice Challenges 'No Surprises': Activist Fights for Right to Revise Plans that Fail to Protect Imperiled Species." *Seattle Post-Intelligencer*, May 4, 2005. http://seattlepi.nwsource.com/specials/licensetokill/222393_leeona04.html (accessed August 1, 2005).

McGivney, Annette. "Moses or Menace?" *Backpacker*, February, 2003.

Medbery, Mike. "Who Knows Best: Grassroots or Foundations?" *High Country News*, October 16, 1995. http://www.hcn.org/servlets/hcn.Article?article_id=1380 (accessed March 1, 2005).

Michels, Robert. *Political Parties: A Sociological Study of the Oligarchical Tendencies of Modern Democracies*. New York: Free Press, 1962.

Mitchell, Robert. "From Conservation to Environmental Movement: The Development of the Modern Environmental Lobbies." In *Government and Environmental Politics: Essays on Historical Developments since World War Two*, edited by Michael Lacey, 81–113. Washington, DC: Wilson Center Press, 1989.

Mitchell, Robert, Angela Mertig, and Riley Dunlap. "Twenty Years of Environmental Mobilization: Trends among National Environmental Organizations." In *American Environmentalism: The U.S. Environmental Movement, 1970–1990*, edited by Riley Dunlap and Angela Mertig, 11–26. Washington, DC: Taylor and Francis, 1992.

Murphy, Michael. "Timber Firms Endangered by Owl Ruling." *Phoenix Gazette*, August 26, 1995.

Noah, Timothy. "Caught in a Trap: Democrats Get Snared by GOP Pact on List of Endangered Species." *Wall Street Journal*, February 17, 1995.

Orr, David. "Reforming the Sierra Club. . . ." *Wild Earth*, Summer, 1991.

Pace, Felice. "Green Myopia." *CounterPunch*, December 19–21, 2008. http://www.counterpunch.org/pace12192008.html (accessed December 31, 2008).

Passoff, Michael. "Shareholder Activism Goes Green." *Earth Island Journal*, Summer, 2001.

Petersen, Shannon. *Acting for Endangered Species: The Statutory Ark*. Lawrence: University Press of Kansas, 2002.

Pickett, Karen. "Of Ballot Initiatives, Political Deals, Millionaires and Ancient Trees." *Terrain*, February, 1992.

———. "Headwaters Activist Killed." *Earth First! Journal* 19(1), 1998.

———. "Redwood and Steel: Steelworkers Join Ranks with Forest Activists." *Earth Island Journal*, Summer, 1999.

———. "Labor and Environmentalists: A Marriage Made in Heaven or a Marriage of Convenience?" *Earth First! Journal* 19(7), 1999.

———. "Is Environmentalism Dead?: A Response to the Recent Essay 'The Death of Environmentalism.'" *Earth Island Journal*, Summer, 2005. http://www.earthisland.org/eijournal/new_articles.cfm?articleID=957&journalID=83 (accessed March 1, 2006).

Piven, Frances, and Richard Cloward. *Poor People's Movements: Why They Succeed, How They Fail*. New York: Vintage, 1977.

Plater, Zygmunt. "Endangered Species Act Lessons over 30 Years and the Legacy of the Snail Darter, a Small Fish in a Pork Barrel." *Environmental Law* 34(2) (2004): 289–308.

Polletta, Francesca. *Freedom Is an Endless Meeting: Democracy in American Social Movements*. Chicago: University of Chicago Press, 2002.

Power, Thomas. *Lost Landscapes and Failed Economies: The Search for a Value of Place*. Washington, DC: Island Press, 1996.

Reffalt, William. "Endangered Species Lists: Chronicles of Extinction?" In *Balancing on the Brink of Extinction: The Endangered Species Act and Lessons for the Future*, edited by Kathryn Kohm, 77–85. Washington, DC: Island Press, 1991.

Roemer, John. "Attorney to the Tree," *San Francisco Daily Journal*, July 22, 1994.

Rogers, Marion. *Acorn Days: The Environmental Defense Fund and How It Grew*. New York: Environmental Defense Fund, 1990.

Rosebraugh, Craig. *Burning Rage of a Dying Planet: Speaking for the Earth Liberation Front*. New York: Lantern, 2004.

Rozek, Victor. "But I'm Just One Person, What Can I Do?" *Forest Voice*, Spring, 1996.

St. Clair, Jeffrey. *Been Brown So Long It Looked Like Green to Me: The Politics of Nature*. Monroe: Common Courage Press, 2004.

———. *Born under a Bad Sky: Notes from the Dark Side of the Earth*. Petrolia: Counter Punch Press and AK Press, 2008.

St. Clair, Jeffrey, and Alexander Cockburn. "Whither the Sierra Club?" *Nature and Politics*, May 26, 1999. http://eatthestate.org/03-36/NaturePolitics.htm (accessed March 1, 2006).

Salladay, Robert. "Biologists Butted in to Save Redwoods." *San Francisco Examiner*, March 2, 1999. http://www.sfgate.com/cgi-bin/article.cgi?f=/e/a/1999/03/02/NEWS12226.dtl&hw=Headwaters&sn=011&sc=211 (accessed March 1, 2005).

———. "Tree Hugger's Hunger Strike at 40 days: Trying to Save Old-Growth Forests, Activist Feeds only on Protest." *San Francisco Chronicle*, November 15, 2002.

Salter, Stephanie. "Attack on Luna Another Test for Hill." *San Francisco Chronicle*, December 3, 2000.

Scarce, Rik. *Eco-Warriors: Understanding the Radical Environmental Movement*. Chicago: Noble Press, 1990.

Seattle Audubon Society v. Evans. 771 F. Supp. 1081 (W.D. Wash.), 1991.

Seideman, David. *Showdown at Opal Creek: The Battle for America's Last Wilderness*. New York: Carroll and Graf Publishers, 1993.

Shellenberger, Michael, and Ted Nordhaus. "The Death of Environmentalism: Global Warming Politics in a Post-Environmental World." Breakthrough Institute. http://www .thebreakthrough.org/images/Death_of_Environmentalism.pdf (accessed February 1, 2006).

Sims, Hank. "Case Dismissed: The Unraveling of People v. Pacific Lumber and People v. Debi August." *North Coast Journal Weekly*, July 7, 2005. http://www.northcoastjournal.com /070705/cover0707.html (accessed August 1, 2005).

Skow, John. "Redwoods: The Last Stand." *Time*, June 6, 1994.

———. "Scorching the Earth to Save It." *Outside*, April, 1999.

Southwest Center for Biological Diversity. *Southwest Center for Biological Diversity Annual Report 1996*. Tucson: Southwest Center for Biological Diversity, 1996.

———. *Southwest Center for Biological Diversity Annual Report 1997*. Tucson: Southwest Center for Biological Diversity, 1997.

Spring and Anne Arcky [pseud.]. "Owl Creek Rises: Confrontation in the Headwaters Complex." *Earth First! Journal* 16(8), 1996.

Stanford Environmental Law Society. *The Endangered Species Act: A Stanford Environmental Law Society Handbook*. Stanford: Stanford University Press, 2001.

Strickler, Karyn. "The Do Nothing Strategy: An Expose of National Progressive Politics." Common Dreams. http://www.commondreams.org/views03/0630-06.htm (accessed April 1, 2005).

Suckling, Kieran. "Some Thoughts on Big Money and Bad Activism." *Earth First! Journal* 16(5), 1996.

———. *Environmental Media Generation Statistics: 2006–2008*. Center for Biological Diversity, Tucson, AZ, 2008.

———. "Ken Salazar a Disappointing Choice for Secretary of the Interior: Strong, More Scientifically Based Leadership Need to Fix Crisis-Plagued Agency." Center for Biological Diversity press release, December 16, 2008. http://www.biologicaldiversity.org/news/press _releases/2008/salazar-12-16-2008.html (accessed Dec 31, 2008).

Suckling, Kieran, and Martin Taylor. "Critical Habitat and Recovery." In *The Endangered Species Act at Thirty*. Vol. 1, *Renewing the Conservation Promise*, edited by Dale Goble, J. Michael Scott, and Frank Davis, 75–89. Washington, DC: Island Press, 2006.

Tahoma [pseud.]. "Cascadia Rising: The Warner Creek Story." *Earth First! Journal* 21(1), 2000.

Takacs, David. *The Idea of Biodiversity: Philosophies of Paradise*. Baltimore: Johns Hopkins University Press, 1996.

Taylor, Ellen. "Corporate Irresponsibility United Us: The Short but Vivid History of the Alliance." *Alliance for Sustainable Jobs and the Environment* 1(1), 2000.

Taylor, Martin, Kieran Suckling, and Jeffrey Rachlinski. "Critical Habitat Significantly Enhances Endangered Species Recovery: Analysis of the three most recent U. S. Fish and Wildlife Service Biennial Reports to Congress on the Recovery of Threatened and Endangered Species." Center for Biological Diversity. www.biologicaldiversity.org/swcbd /programs/policy/ch/index.html (accessed December 1, 2004).

Thomas, C., A. Cameron, R. Green, M. Bakkenes, L. Beaumont, Y. Collingham, B. Erasmus, et al. "Extinction Risk from Climate Change," *Nature* 427 (2004): 145–48.

Thompson, A. Clay. "Baked Goods: Our Reporter Enters the Shadowy World of the Biotic Baking Brigade." *San Francisco Bay Guardian*, December 9, 1998. http://www.sfbg.com /News/33/10/Features/bbb.html (accessed March 1, 2005).

Tobin, Richard. *The Expendable Future: U.S. Politics and the Protection of Biological Diversity*. Durham: Duke University Press, 1990.

Tokar, Brian. *Earth for Sale: Reclaiming Ecology in the Age of Corporate Greenwash*. Boston: South End Press, 1997.

U.S. Department of the Interior Office of Inspector General. *Audit Report: The Endangered Species Program, U.S. Fish and Wildlife Service*. Report No. 90-98, 1990.

Vaughn, Jacqueline, and Hanna Cortner. *George W. Bush's Healthy Forests: Reframing the Environmental Debate*. Boulder: University Press of Colorado, 2005.

Walley, J. Zane. "Playing Outside the Rules." *Range*, Winter, 1999. http://www.rangemagazine .com/archives/stories/winter99/playing_outside_the_rules.htm (accessed March 1, 2005).

Walters, Mark. "California's Chain-Saw Massacre." *Reader's Digest*, November, 1989.

Watson, Paul. "Tora! Tora! Tora!" In *The Earth First Reader: Ten Years of Radical Environmentalism*, edited by John Davis, 33–42. Salt Lake City: Peregrine Smith, 1991.

Wayburn, Laurie, Jerry Franklin, John Gordon, Clark Binkley, David Mladenoff, and Norman Christensen Jr. *Forest Carbon in the United States: Opportunities and Options for Private Lands*. San Francisco: Pacific Forest Trust, 2007.

Weber, Michael. *From Abundance to Scarcity: A History of U.S. Marine Fisheries Policy*. Washington, DC: Island Press, 2002.

"What Happened That Day: Accounts from Earth First!" *Earth First! Journal* 19(1), 1998.

Wilkerson, Matt. "Climate Activists Invade DC Offices of Environmental Defense Fund: Daughter of Environmental Defense Fund Founder Accuses NGO of Pushing False Solutions to Climate Change." *Rachel's Democracy and Health News*, #988, December 4, 2008. http://www.precaution.org/lib/08/ht081204.htm#Climate_Activists_Invade_D.C._Offices _of_Environmental_Defense_Fund (accessed Dec 5, 2008).

Wilson, Janet. "Enforcement of Species Act Pits Babbitt vs. Activists." *Los Angeles Times*, October 3, 1999.

Wreckin' Ball [pseud.]. "Union/Eco Alliance Costs Hurwitz Half a Mil." *Earth First! Journal* 18(8), 1999.

Wuerthner, George. ed. *Wildfire: A Century of Failed Forest Policy*. Washington, DC: Island Press, 2006.

Yaffee, Steven. *Prohibitive Policy: Implementing the Federal Endangered Species Act*. Cambridge: MIT Press, 1982.

———. *The Wisdom of the Spotted Owl: Policy Lessons for a New Century*. Washington, DC: Island Press, 1994.

Young, Buck [pseud.]. "Abracadabra!: Defending the Wild with Magic (and Lots of Ink)." *Earth First! Journal* 21(1), 2000.

Young, Margaret. "Nightmare on Polk Street: ASCMEE Acts Up." *Wild Earth*, Winter, 1991.

———. "Wilderness Now!" *Wild Earth*, Spring, 1993.

Zakin, Susan. *Coyotes and Town Dogs: Earth First! and the Environmental Movement*. New York: Viking Books, 1993.

———. "The Wild Bunch: Why Does the Center for Biological Diversity Keep Winning?" *Tucson Weekly*, June 21, 2001. http://www.tucsonweekly.com/gbase/Opinion/Content?oid =44163 (accessed March 1, 2005).

INDEX

University of Oregon environmental law
  conference, 93
Urban Habitat, 254n58

V

Van Dyck, Tom, 60-61, 69, 250n32
Voice of the Environment, 55
Voss, Rene, 140, 141, 144, 254n62, 255n74

W

Warner Creek blockade, 133-34
Water quality, 102-3
Water Resources Control Board, 102-3
Watt, James, 26, 167, 169, 202
Werbach, Adam, 12, 247n22, 254n66
Western Ancient Forest Campaign, 255n74
Wilburforce Foundation, 253n38, 261n110
Wild Alabama, 5, 34, 113, 125, 158
Wild Earth, 129, 218
WildEarth Guardians, 240. *See also* Forest
  Guardians.
Wilderness Act (1964), 18-19, 26, 141
Wilderness legislation, 19, 114-15, 262n13
Wilderness Society
  ESA and, 172
  Health Forest Initiative and, 152-53
  media and, 83, 230
  opposition to NFC legislation, 127
  origins of, 16
  OWC and, 115
  spotted owl and, 120, 158
  United Forest Defense Campaign and, 154
  Wilderness Act and, 18-19
Wilson, Pete, 82
Wolke, Howie, 28
Women, position in grassroots groups of, 46,
  252n77, 260n108
Wood supply, 112, 139, 140

World Trade Organization (WTO) meeting,
  Seattle, 94-95
World Wildlife Fund, 190, 210, 242, 247n1,
  259n78
Wyden, Ron, 153

Y

Yaffee, Steven, 162, 166, 243
Yannacone, Victor, 20, 209-10
Young, Margaret Hays, 128, 129, 253n46

Z

Zakin, Susan, 116
Zero-cut campaign
  overview, 7-8, 112-13
  and (George W.) Bush administration, 151-155
  economic costs of national forest logging,
    124, 139-40
  "end commercial logging," 151-52, 254n57
  funding and, 146-47, 154, 219
  Heartwood, 124-26, 130
  impact of, 155-59
  John Muir Project, 138-41
  John Muir Sierrans, 130-31, 134-38
  litigation and appeals, 145-46, 147, 157
  and logging levels, 112f, 155, 158
  National Forest Protection Alliance, 147-48
  National Forest Protection and Restoration
    Act (NFPRA), 140-42
  national forests, role of, 112
  nationals, opposition of, 112-13, 156-58
  Native Forest Council, 126-28
  Protect Our Woods (POW), 123-24
  pubic opinion survey results, 148, 246n13
  relation to Roadless Rule and the Heritage
    Forests Campaign, 148-50, 229
  timber workers and, 139
  wood supply and, 112, 139, 140

ABOUT THE AUTHOR

Douglas Bevington is the forest program director for Environment Now, a grantmaking foundation based in California. He received his PhD in sociology from the University of California, Santa Cruz, where he taught courses on social movement studies. He has been involved in forest and wildlife protection activism for more than twenty years.